Ambrose Loomis Ranney, Ambrose L. (Ambrose Loomis) Ranney

Eye-strain in health and disease

with speical reference to the amelioration or cure of chronic nervous

derangements without the aid of drugs

Ambrose Loomis Ranney, Ambrose L. (Ambrose Loomis) Ranney

Eye-strain in health and disease
with speical reference to the amelioration or cure of chronic nervous derangements without the aid of drugs

ISBN/EAN: 9783744735988

Printed in Europe, USA, Canada, Australia, Japan

Cover: Foto ©berggeist007 / pixelio.de

More available books at **www.hansebooks.com**

EYE-STRAIN

IN HEALTH AND DISEASE.

WITH SPECIAL REFERENCE TO THE AMELIORATION OR CURE
OF CHRONIC NERVOUS DERANGEMENTS WITHOUT
THE AID OF DRUGS.

BY

AMBROSE L. RANNEY, A.M., M.D.,

AUTHOR OF "LECTURES ON NERVOUS DISEASES," "THE APPLIED ANATOMY OF THE NERVOUS SYSTEM," "A
TREATISE ON SURGICAL DIAGNOSIS," "PRACTICAL MEDICAL ANATOMY," ETC.; LATE PROFESSOR OF THE
ANATOMY OF THE NERVOUS SYSTEM IN THE NEW YORK POST-GRADUATE MEDICAL SCHOOL
AND HOSPITAL; LATE PROFESSOR OF NERVOUS DISEASES IN THE MEDICAL
DEPARTMENT OF THE UNIVERSITY OF VERMONT, ETC.

Illustrated with Thirty-Eight Wood-Engravings.

PHILADELPHIA, NEW YORK, CHICAGO:
THE F. A. DAVIS COMPANY, PUBLISHERS.
1897.

PREFACE.

—

THIS volume comprises the substance of several mono-graphs that the author has published from time to time during the past ten years in medical journals, with the addition of considerable new matter. He has added, also, the histories of many typical cases in detail with the view of illustrating some remarkable results of eye-treatment alone upon various forms of nervous disturbances that have persisted for years and failed to yield to any other form of treatment.

Many of the histories published in this volume are given with sufficient completeness to shed much light upon the methods employed in each case, as well to demonstrate the results obtained by the use of glasses and graduated tenotomies upon some of the ocular muscles.[1] To the oculist the technical portion of some of these records will doubtless prove of greater interest than to the general practitioner in medicine; but the author trusts that the labor involved in preparing such histories for the press from scattered office memoranda will not be deemed by any reader as misspent.

The views which the author supported in his work on nervous diseases[2]—relative to the effects of eye-strain upon the development of headache, neuralgia, sleeplessness, chorea, epilepsy, nervous prostration, and insanity—are reiterated here with strong clinical evidence to sustain them. Time has but strengthened the author's early convictions, while many of those who were antagonistic to these views years ago are now enthusiastic in their support.

In Chapter II of this work an attempt has been made to so present the various steps that should be taken during an examination of an eye for errors of refraction, and of the ocular muscles for anomalies of adjustment of the eyes (heterophoria), as to enable even a beginner in this work to come to some

---

[1] This is particularly true of the epileptic cases (Chapter VII).
[2] Lectures on Nervous Diseases, The F. A. Davis Co., Philadelphia, 1888.

definite conclusions regarding the presence or absence of eye-strain without aid from a specialist.

To the critics the author would say that three facts should not be lost sight of in this volume,—viz., that none of the cases here reported took any drugs while under his care, that they were chronic cases which had received no benefits from medication under skillful hands, and that many of them were made absolutely well by eye-treatment alone.

To appreciate the fact that, within the past ten years, a remarkable change has occurred in reference to this important field, it is only necessary to contrast the later works with those of earlier date upon diseases of the eye and diseases of the nervous system. Few, if any, can be found to-day that have ignored entirely the relationship between eye-strain and functional neuroses, while in the past this field was neither hinted at nor discussed in any of the text-books. Moreover, the medical journals to-day are constantly furnished with interesting articles relating to this subject both in America and Europe. Frequent and long discussions in societies regarding eye-muscles, the new instruments that are being devised, and the growing sale of these instruments also point toward the rapidly increasing interest in this field of medical work.

While the author does not expect that his critics will accept all of his conclusions, he has a right to ask that the work be read without prejudice prior to criticism and that the reviews be dispassionate. The time has passed when violent antagonism or ridicule can have much weight in opposing the progress of any method of treatment of disease that is scientific and positive rather than speculative or empirical.

The author would express here his indebtedness to his assistant, Dr. Wm. R. Broughton, for valuable aid in preparing for the press many of the clinical histories that are incorporated in this volume, the tabular matter, and the index.

AMBROSE L. RANNEY.

345 MADISON AVENUE, NEW YORK CITY,
June, 1897.

TABLE OF CONTENTS.

CHAPTER I.

The underlying factors of disease.
Eye-strain as a cause of excessive nervous expenditure.
Modern methods of testing for anomalies of the eye-muscles.
The far-sighted eye and its effects upon health.
The near-sighted eye and its effects upon health.
The astigmatic eye and its effects upon health.
The heterophoric eye and its effects upon health.
Hereditary predisposition to disease and its problems.
 The so-called "tubercular" tendency.
 The so-called "gouty" tendency.
General summary and conclusions.

CHAPTER II.

(a) Tests for refraction.
 Suspected hypermetropia.
 Suspected myopia.
 Suspected astigmatism.
 Methods of recording refractive tests.
(b) Tests for heterophoria.
 Movement-test.
 Exclusion-test.
 Red-glass test.
 Phorometer-test for esophoria.
 Phorometer-test for exophoria.
 Phorometer-test for hyperphoria.
 Maddox's rod-test.
 Savage's rod-test.
 Stevens's lens-test.
 Measuring the strength of individual muscles.
  Power of adduction.
  Power of abduction.
  Power of sursumduction.
 Tests within range of accommodation.
 Methods of recording tests for heterophoria.
(c) Atropine tests.
 Importance of, in determining refractive errors.
 The uncertainties of the ophthalmoscope.
(d) Investigation of latent heterophoria by prisms.
 Special outfit.
 General rules.
(e) Bilateral deviation of eyes in the vertical meridian.
 Tests for "anaphoria" and "anatropia."
 Tests for "kataphoria" and "katatropia."
 Stevens's tropometer.
(f) Measurement and recording of visual field.
 The perimeter.
 The normal field.
(g) Diagnostic attitudes of the head and peculiar facial expressions caused
  by heterophoria.
 Photographic illustrations.
(h) The ophthalmoscope.
 Special indications for its use.
 Its unreliability in determining refraction in certain cases.
 General rules for its employment.

CHAPTER III.

General classification of headaches and neuralgias.
Toxic headaches and neuralgias.
From organic lesions of brain, spinal cord, or nerves.
Reflex headaches and neuralgias.
Popular fallacies regarding the causes and treatment of headaches and neu-
ralgias.
Hypermetropia as a very frequent, but seldom recognized, cause.
Asthenopic headaches and neuralgias.
Heterophoric headaches and neuralgias.
Clinical histories of illustrative cases and a tabulated summary.

CHAPTER IV.

The various theories of to-day regarding the causation of chorea.
Microbe theory.
Toxic theory.
Rheumatic theory.
Fallacies of visionary speculation.
The reflex factors in causation.
Medical support of this view.
Direct and positive clinical evidence.
General conclusions regarding chorea.
Enormous percentage of uncorrected hypermetropia.
Imperfectly fitted frames, even when glasses have been prescribed.
Latent heterophoria very often present, but unrecognized.
Prismatic glasses not curative, as a rule.
Clinical histories of illustrative cases.

CHAPTER V.

Frequency of this condition and its results.
Most common among the brain-workers.
Generally a purely reflex symptom.
Physiological explanation of how eye-strain can cause sleeplessness.
Reasons why the eye-treatment may have failed to relieve.
Eye-records often incomplete.
Old methods too often employed.
Conclusions too hastily arrived at.
Clinical histories of illustrative cases.

CHAPTER VI.

Are the causes of these conditions yet fully understood?
Exhausted nervous vitality usually precedes and accompanies digestive ills.
Machinery will not run without proportionate power.
Can eye-strain alone cause disturbances of the stomach, liver, and intestine?
The effects of a persistent leakage of nerve-force upon the various organs.
May not obstinate constipation be simply a symptom of defective nerve-
power?
Intense gastric pain and vomiting is frequently caused by reflex disturb-
ances from the eyes.
Clinical histories of illustrative cases.

CHAPTER IX.

CHAPTER X.

Eye-Strain in Health and Disease.

CHAPTER I.

THE BEARINGS OF EYE-STRAIN UPON THE DURATION OF HUMAN LIFE.

To counteract the *underlying factors of disease* is even more important than to combat disease when actually developed.

Life is, at best, precarious. "Three score years and ten" are reached but by a few. From infancy we are exposed to conditions which either predispose toward or actually excite disease. The basis and aim of all life-insurance examinations is to determine, as far as human skill can, the existence or non-existence of two factors which necessarily modify the risk to life. These are as follow :—

1. *Whether the applicant, by heredity, climatic surroundings, occupation, personal habits, and other causes, is not more than ordinarily liable to the development of disease.*

2. *Whether any evidence of existing weakness or of actual disease can be discovered by a physical examination.*

As examples of the former of these two factors, let me cite a few familiar illustrations which not infrequently present themselves. An applicant, for instance, who shows an unmistakable family tendency to consumption, and whose ancestors have seldom passed the age of 45 years, would not at the age of 30 (even if in perfect health) be considered as good a risk as if his ancestry had shown no evidence of a phthisical tendency and had been long-lived.

Again, one who from necessity was a resident of a malarial region, or where periodical epidemics of fevers were common, would be materially affected as a risk by that fact, in spite of good health.

Hazardous occupations increase the risk to life. A confirmed habit of excessive indulgence in alcohol does the same. A victim to the use of drugs, such as the morphine, chloral, or cocaine habits (which slowly tend to sap the nervous vitality of the patient, and to render him, as a consequence, more than ordinarily susceptible to atmospheric germs and disease), would be rejected by any careful medical examiner.

It is well and justly recognized to-day by the community at large that it is necessary, for the protection of the interests of the insured, as well as that of the insurer, that every step be taken to ascertain, and every safeguard be used to overcome, through the medical examiners, upon whose judgment the risk is to be accepted or rejected, all *predisposing factors* to the development of disease in any applicant for insurance.

The protection of the widow and the fatherless rests upon the faith of the community, at the present day, in the competency of the medical men, who, by their skill and discriminating judgment, keep the death-rate from assuming proportions that would prove hazardous to the financial prosperity of the corporations that have diversified financial interests, involving millions, in their keeping.

I desire, therefore, that the thoughtful reader of these pages shall see in them no affectation of science or pretence to superior knowledge, but an honest and earnest purpose to bring to their notice, in a manner that is simple and practical, a statement of clinical facts that are to-day receiving no small amount of attention from scientific medical men, and a theory that is based upon nature and common sense.

In endeavoring to present the view (now quite generally accepted) that the eyes themselves may (when defective in refraction or when imperfectly adjusted so that they fail to work in harmony with each other) constitute an *important and too commonly neglected factor*, both in *causing and perpetuating disease*, I believe and trust that I shall open to the minds of some of my readers a field worthy of serious thought and careful consideration.

Within the past few years the attention of the medical profession has been drawn, more forcibly than ever before, to the fact that eye-strain may constitute an important element in the causation of all nervous disturbances of the so-called " functional " type, and also of many symptoms referable to the viscera. The latter are too often construed as indications of actual disease of the organ disturbed.

In the light shed upon this subject, chiefly by recent contributions to medical literature, the view is gradually being accepted by many in the profession that certain nervous diseases (whose pathology, to say the least, is still in doubt) are possibly not dependent, in every case, upon an unrecognized organic lesion ; and they are being led to coincide with the statement that the term "*functional*" nervous disease may be properly applied, in some instances, at least, to the graver nervous conditions,—such, for example, as epilepsy, chorea, hysteria, or other manifestations of nervous exhaustion, and insanity.

In other words, the professional mind seems more willing now than in the past to discard an apparently fruitless search for a pathognomonic lesion for each intractable nervous condition, and to look more calmly upon tangible clinical facts, even if they are radically opposed to pre-existing views.

CAN NERVOUS DISEASES BE CAUSED BY EYE-STRAIN?

The literature of medicine goes to prove conclusively that the duration of life is materially shortened by nervous debility and the diseases which it entails. Any factor, therefore, in their causation ought not to be overlooked. This subject of inquiry has become invested with an importance which cannot well be ignored by searchers after truth.

If the view that *eye-strain is a frequent cause of functional nervous derangements* proves to be the correct one, beyond the possibility of a doubt or cavil, it is not difficult to see that a hope of marked relief or of ultimate recovery is practically extended to many hopeless sufferers upon whom drugs have exerted little or no benefit.

In anticipation of the general acceptance of such a view (which, I feel assured, must, in time, prevail), I have deemed it wise to first discuss, from a physiological stand-point, some of the points involved in this theory, and, later, to illustrate, by selected cases, the curative effects of eye-treatment upon the various types of nervous derangement.

In order that those of my readers (who have, possibly, not given much attention to the views which these cases are particularly selected to illustrate) may properly understand the train of reasoning that offered a solution, to my mind, of the symptoms recorded, I take the liberty of quoting a few paragraphs from a paper which I read before the International Medical

Congress at Washington, entitled " Does a Relationship Exist between Anomalies of the Visual Apparatus and the So-called ' Neuropathic ' Predisposition ? " [1]   This paper was based upon a carefully tabulated analysis of the records of one hundred consecutive cases of typical neuroses taken from my private case-book.

In this paper I say: " Until there is a uniformity in the methods employed for testing the eye-muscles, and of terms for the recording of anomalies so detected, the profession must unfortunately continue to be more or less embarrassed in this line of research.   I do not feel justified in personally discussing this subject here, as it has only an indirect relationship with this paper; but I cannot refrain from saying, in this connection, that to defective methods of examination, made venerable chiefly by their antiquity, we owe to-day, in my opinion, much of our ignorance of anomalies of the ocular muscles."

Some time ago I was struck, on looking over a children's magazine, with an illustration designed to teach the reader the dependence of the various organs of the body upon the brain. It represented the brain as the head of a manufacturing establishment sitting at his desk, and around him were the various departments,—as, for example, the liver-department, the stomach-department, the eye-department, etc.   These departments were connected with the head of the establishment (the brain) by telegraph-wires, through which each could make its wants known and receive information regarding them.

Probably the designer of this sketch (made for the purpose of illustrating to the child the dependence of the organs upon the brain for their successful operation, as well as their actual support) built " better than he knew."   He embodied in his drawing a graphic representation of certain fundamental principles of physiology which are not clearly understood, even by many adult minds, in their bearings upon the general health.

The lungs do not make us breathe, except in an indirect way,—by asking the brain to start the necessary muscles into action.   The stomach does not perform its functions until after the brain has been requested by it to turn on the blood-supply in sufficient quantities to produce the requisite amount of gastric juice.   The intestine performs its incessant, worm-like movements by no inherent power of its own.   The heart keeps

---

[1] An abstract of this paper was published in the Medical Register, November 19, 1887.

up its rhythmical beating only when permitted to do so by the great center of nerve-force.

Now, is it at all inconsistent with physiological principles to advance the view that *any excess of nervous expenditure to one organ over the normal amount which should be furnished is done at the expense of the others sooner or later?*

No one can draw incessantly upon his reserve-capital of nerve-force without incurring a risk of ultimately exhausting it. A *bankruptcy in the reserve-capital of nerve-force* entails untold ills to the individual.

The day of reckoning is postponed in any given case in direct proportion to the drafts made upon the reserve and the amount of the reserve. This may help us to explain why some escape it indefinitely while others are precipitated into indescribable distress when life is hardly begun.

In case the bearing of eye-strain upon the problem of nervous expenditure is not very clear to some of my readers, I deem it wise to call attention to some facts relating to the more common ocular defects that are capable of transmission from parents to their offspring.

Although something has been written within the past few years in relation to the deleterious effects of errors of refraction and accommodation of vision and the condition known as "*muscular insufficiency*" upon the functions of the *nervous system and the viscera*, the profession at large is not yet thoroughly awakened to the importance of the detection and correction of such errors.

Most men know that some persons can be made dizzy by looking from a height or inspecting a water-fall; they have doubtless seen people suffer pains in the head and be made "sick at the stomach" by trying on a pair of spectacles which gave relief to a friend. All physicians doubtless know that a "squint" in the eye is very often due to some defect in the refraction of the eye or a weakness of its muscles; but possibly some of them do not know that a squint will occasionally disappear at once, when the proper glasses are given to such a patient, without recourse to cutting the muscle. Perhaps it has never occurred to most of my readers that sight is the *only special sense which we use constantly,* except during the hours of sleep. There is not a moment of the day when we are not acquiring visual impressions of some kind.

Fortunately for our nervous system, the normal eye takes pictures of surrounding objects *without any muscular effort* when the object is more than twenty feet away; hence, during the larger part of each day, the *normal eye is passive,* and is practically at rest, although performing its functions.

How different is the condition of the *far-sighted,* or "*hypermetropic*," eye, however, from the normal! For this eye (since it is *too short* in its antero-posterior axis) all objects *have to be focused by muscular effort,* irrespective of their distance from the eye. Such an eye is never passive. It has no rest while the body is awake. It is always straining, more or less intensely, to bring properly upon the retina the images of objects seen.

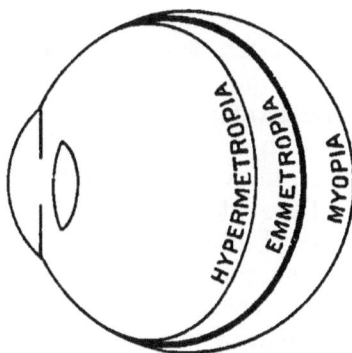

FIG. 1.—A DIAGRAM DESIGNED TO ILLUSTRATE CONGENITAL OR ACQUIRED MALFORMATIONS IN THE ANTERO-POSTERIOR AXIS OF THE EYEBALL.

THE SHALLOW EYE.—The "*hypermetropic*" condition of the eye, or "*far-sightedness,*" as it is called, is a very common defect. It is especially frequent in persons of tubercular parentage.[1] It is well, therefore, to suspect the existence of this defect in children or adults whose ancestors have died of "consumption."

Hypermetropia cannot be corrected too early in life. It is unquestionably one of the *most frequent causes of "sick headache,"* which, as we all know, runs in families. It is commonly encountered also (among other optical defects) in subjects afflicted with chorea and epilepsy. It is a congenital defect, and will never be "outgrown," as many people think. A hypermetropic child, from the days of babyhood, suffers (un-

---

[1] This is probably due to the shallowness of the orbits.

conscious, perhaps, of the fact) from a variety of symptoms which indicate the "strain" to which it is subjected in consequence of its efforts to see distinctly. Its eyes are likely to become easily suffused when it plays or looks steadily at near objects. A slight cast in the eye is sometimes developed. It occasionally "sees double" after it learns to read. It usually prefers and excels in out-of-door sports, which require only slight efforts at accommodation of vision. It finds that study and close application to books bring an indescribable sense of weariness and discomfort; hence study becomes irksome and play brings a sense of peculiar relief.

Now, one clinical fact should be noted here,—viz., that *hypermetropic subjects often have remarkable acuteness of sight.* They are very apt (when young adults) to boast of their power of vision. They can often read all the test-types made for distance (twenty feet or more) without an error. If the defect exists in a child, the parents will frequently tell you how the child can see things with distinctness which possibly they themselves cannot see at all; how they have tested its eyes from time to time; how absurd the idea seems to them and their friends that the vision of the child is defective; and how unnecessary the use of glasses seems to them (even if the eye is abnormal) so long as the child can get along without them. In some cases no amount of explanation or pleading will persuade the parents to have the ophthalmoscope or atropine used upon the child's eyes in order to decide the question of the existence of "*latent*" far-sightedness.

Some years ago I pleaded with a medical man to allow some oculist of reputation to examine his children's eyes, all of whom had weekly attacks of sick headache, inherited from both the mother and father, and in whom a tubercular tendency was strongly marked. I was refused, and the statement was made that never, while the father lived, should a child of his wear glasses with his consent. One of these children wears to-day a convex glass with a twelve-inch focus for distance; another wears the same glass with five degrees of prisms added. These only partially correct an insufficiency of the muscles, which exists in addition to the hypermetropia. A third child is highly hypermetropic and astigmatic. In every one of these subjects immense relief has been afforded by the correction of an optical defect which had rendered their early life one of suffering. This is not an uncommon experience. I could cite many more, if I deemed it necessary to prove what is already accepted by ophthalmologists as demonstrated,—viz., that hypermetropia and eye-defect of other forms may become *fruitful sources of headache.*

There is a prejudice among laymen and some medical men that glasses are an injury when they can be avoided, because, as they say. "a person becomes so dependent upon them when he once puts them on." This argument should be exactly reversed, and construed as follows : "*Because nature becomes dependent upon a glass which gives relief and corrects an existing strain upon the eye, no time should be lost in affording this relief.*"

Should a hip-splint be avoided (when the pain in the joint is arrested by it) because the patient feels his dependence upon the splint? Should a child be allowed to go through life with a deformed eye simply because the defect is not apparent to himself or his friends on account of an unnaturally developed ciliary muscle, which for a time renders the eye capable of getting along tolerably well in spite of its deformity?

More harm is being done to-day to the community at large by this fallacious argument than is possible to compute. Thousands of sufferers from sick headache and neuralgia are to-day struggling along through life with an optical defect uncorrected, and, in many instances, after costly experimentation with drugs and doctors, are left in despair of cure.

I speak strongly upon this point, because I believe that the gastric symptoms which accompany typical attacks of sick headache are not to be explained (as they commonly are) on the ground that the "liver is inactive." or that "dyspepsia exists," or that the "gastric juice is weak," or that "the patient uses tobacco to excess." or that "he has been living too high." Every one who has suffered for years with these attacks knows that they often occur without explainable cause; that they are cured sometimes by eating, drinking, and smoking, and made worse at other times by similar indulgences or excesses; that every known remedy is apt, sooner or later, to prove inoperative, and that a sure specific for them is unknown among the drugs of our pharmacopœia. These subjects also know that life is rendered almost unendurable by the attacks at times. They are tractable patients, and will try anything, live in any way specified, and bear any privation without a murmur, if it will insure a cure.

I believe, from a personal experience of my own of this kind (which it is unnecessary to relate here), and from some experience also in examining the eyes of this class of sufferers, that the symptoms of sick headache are reflex in character to a large extent, and are due, primarily, in almost every case to some optical defect. We can easily demonstrate that disturbed brain-action from eye-strain may produce in a healthy child and in some adults all of the symptoms of these attacks in a few minutes. Why is it irrational, therefore, to affirm that a brain disturbed by the constant efforts made to use eyes, which are

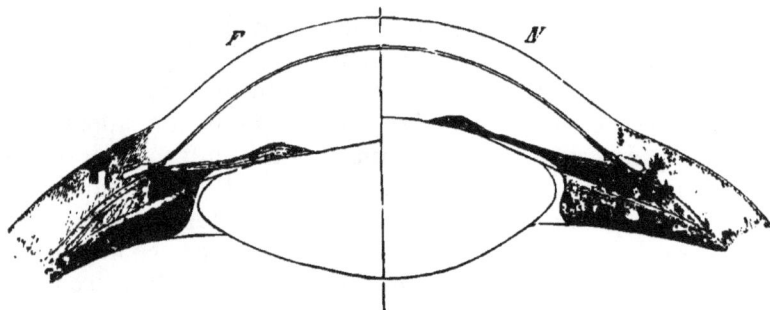

FIG. 2.—SECTION OF THE FRONT PART OF THE EYE, SHOWING THE MECHANISM OF ACCOMMODATION. (Fick.)

The left side of the figure (*F*) shows the lens adapted to vision at *distances of over twenty* feet; the right side of the figure (*N*) shows the lens adapted to the *vision of near objects*, the ciliary muscle being contracted and the suspensory ligament of the lens consequently relaxed.

abnormal in respect to the refraction, accommodation, or the equilibrium which should exist between its various muscles, may manifest its disturbed state by nausea, headache, vomiting, dizziness, constipation, and other evidences of imperfect performance of the functions of the viscera? Does not our central nervous system regulate and directly control those functions? Is it not as probable that the master, when upset, disturbs the servants under him, as to advance the argument that the servants themselves are the all-important factors in the causation?

THE ELONGATED EYE.—A second defect of the eye, which is less liable to cause physical debility than the hypermetropic eye (already described), now merits attention.

When the eye is too long from before backward the patient

is said to be "*myopic*," or *near-sighted.* Distant objects are more or less indistinct to such an eye in proportion to the excessive length of the antero-posterior axis of the eye over the normal standard.

No amount of muscular effort can overcome or improve this defect in vision except through compressing the eyeball slightly by squinting the eyelids; hence these individuals are not subjected to the muscular strain which far-sighted persons constantly and unconsciously exert in order to see at a distance.

Again, the near-sighted eye can read or perform any of the functions required of it (when brought sufficiently close to the object) without any muscular effort of an unnatural character. In contrast, the far-sighted eye has to exert a still greater muscular effort to see near objects distinctly than when employed upon distant objects; hence the fatigue, the blurring of letters upon a printed page, the watering of the eyes, the pain in the eyes and head, and the many other ills previously described.

Near-sighted subjects are generally conscious of an eye-defect, because they cannot see across a room with distinctness or recognize familiar faces on the street. They are apt to become very fond of occupations which bring the eye close to their work, because they have no difficulty in seeing the object. Near-sighted children are liable to be considered precocious beyond their years, because they prefer to read rather than to play out-of-doors. It is generally safe to conclude that a child is near-sighted when it avoids out-of-door amusements in order to gratify a taste for reading or in-door occupations.

Near-sightedness is less liable to induce nervous disturbances than far-sightedness, provided it is not accompanied by astigmatism or muscular insufficiency. Yet it should be remembered that myopic subjects are more frequently sent to the oculist for relief than hypermetropic subjects are, because the defect in vision is very apparent to all in the former class, and is more often unsuspected than recognized in the latter.

Finally, it behooves us to consider, in a general way, the condition known as "*astigmatism*" and also certain anomalies of the muscular adjustment of the eyes, from which harmonious action of the two often becomes materially disturbed and, in some patients, completely overthrown.

THE ASTIGMATIC EYE.—The physician may find, in the third place, when he has examined the eyes of patients or friends who suffer from headache, persistent neuralgic attacks, etc., that a condition of the eye known as "*astigmatism*" may be detected, co-existing with far- or near- sightedness or independent of these refractive errors. In such cases the cornea or the lens of the eye has a *greater curvature in some meridians than in others ;* hence the images of all objects seen are more or less distorted when they fall upon the retina. To this class of sufferers some letters in the tests employed will be distinct, while others will not. If a number of dots are made upon a blackboard or a sheet of paper, some will appear as ovals, with a hazy border, or as lines, while others will more closely resemble the normal appearance of the dots. Finally, if a card, with lines running from its centre to its periphery (the " clock-face test ") is used, some of the lines will appear blacker than the rest and more clearly defined. Now, there can be no comfort to such subjects in their visual efforts.[1] They learn by practice and experience to properly interpret, after awhile, the imperfect images of objects seen, and they are aided in so doing by the fact that the outlines of letters, etc., become clearer in some positions, as regards the eyes, than in others; but, in spite of all that may be said to the contrary, the strain of using imperfect eyes tells upon most astigmatic persons sooner or later, and tends to excite reflex nervous phenomena of various kinds. To properly correct astigmatism by glasses is often an extremely difficult matter. It requires experience, a thorough knowledge of optics, and a familiarity with the practical use of the ophthalmoscope. There are comparatively few physicians (outside of the specialists in ophthalmology) who are capable of managing a bad case of this kind with perfect success. Almost any physician can, however, easily detect its existence in most cases. When he discovers it I would advise him to intrust its correction to skillful hands.[2]

THE HETEROPHORIC EYE.—Finally, it is very important that the physician determine (in each patient whose eyes are examined by him) the *condition of the muscles* of the eye. The term "*heterophoria*" is generally applied to that condition of the visual apparatus which entails suffering in consequence of a

---

[1] This subject is discussed more fully in Chapter II.
[2] Cohn shows that, in 299 eyes under atropine, no case of absolute emmetropia was detected. Hausen found but 26 emmetropic eyes in 1610 and Dürr but 30 in 414. A. Randall states, in his article on " The Refraction of the Human Eye, a Critical Study of the Examination of the Refraction, especially among School-children " (American Journal of the Medical Sciences, July, 1885) that only $7\frac{34}{100}$ per cent. of 1834 eyes of infants and school-children were found to be emmetropic.

defective " equilibrium " in the muscular power exerted upon that organ when a fixed position of the eye is maintained for any length of time. When a state of perfect equilibrium is impaired from either a weakness or abnormal tension of some muscle of the eye, the effects are apt to become manifested sooner or later by pain and great discomfort after the eyes are used for any length of time. I have seen patients who could not attend a place of amusement, read, or sew, for even a short time, without great distress from this cause. These patients may or may not have a refractive error. In some instances no glasses but *prismatic ones* will benefit them.

A high-couraged horse feels the will, as well as the support, of his driver through the reins by means of the bit. Although his course and rate of speed are changed from time to time at the will of the driver, the reins are never slackened. The horse becomes acquainted with the desires of his master by a sense of increased or diminished tension upon the reins. He is guided to either side by a difference in the tension of the two, although the driver does not entirely relax his hold upon the opposing rein while he uses the guiding one, and the difference in tension may be very slight.

So it is with the normal eye. It is both controlled and supported, while performing its movements within the orbit, by the eye-muscles (which are its reins). The brain is the driver. At its command the eye revolves or remains stationary at any desired point. The tension of muscles, opposed to any movement of the eye required, is so modified by the brain as to insure the requisite support to the eyeball, and to steady it as it moves. Thus a perfect equipoise is constantly established between opposing forces, adjusted with the nicest care to meet the full requirements of the organ under all possible circumstances. The normal eye does not tremble or wabble when it moves or the attempt is made to hold it in any fixed attitude. It is a piece of machinery, perfect in all its parts, reliable in its movements, perfectly controlled by its master.

The eye with muscular insufficiency is like a horse with an inexperienced and incompetent driver; the proper tension upon the reins is not maintained at all times, as it should be; there is no equilibrium between antagonistic muscles; fixed attitudes are maintained with difficulty for any length of time; the brain becomes more or less disturbed by its inability to properly control the eye-movements, and exhausted by the continual strain imposed upon it by the efforts required to do so even imperfectly.

Subjects of this class are *very frequently encountered* in the practice of a neurologist. The oculist, perhaps, sees them still oftener, because they are generally conscious that something is wrong with their eyes. Still, there are exceptions to this rule. I have examined patients who showed, in response to appropriate tests, very high degrees of heterophoria (see page 11), that came to me for the relief of symptoms which had never been referred, by themselves or their physician, to any possible eye-defect. I recall the case of an epileptic who was placed under my charge. His family assured me he had " wonderful eyes," and they were surprised when I examined them with care. The results of this examination showed, however, that twenty-five degrees of insufficiency of the externi existed (as measured by the vertical-diplopia test), and that he was hypermetropic and astigmatic to a marked degree.

Insufficiency of ocular muscles seems to me to be a *congenital defect*, in most cases,—possibly in all. It is encountered in very young subjects. It is not a paralysis or a true paresis. It is not uncommon to note wide variations in the same case, if examinations are made from time to time, under certain favorable conditions. Possibly this fact helps to explain why competent observers do not always estimate the degree of heterophoria in a given case alike, even when similar tests are employed and equal care is given to the case. This point is discussed in Chapter II as an evidence of " *latent* " heterophoria.

We have no way, as yet, of determining " *latent heterophoria*" as we do latent hypermetropia by atropine. Should a patient show us an error of adjustment counteracted by a prism of a certain angle to-day, it only proves that he has *at least* that amount, not that he has no more. This statement can. I think, be demonstrated. It is an important fact to remember, when the results of examinations of such patients made by one's self are at variance with the observations made by another.

ARE THE QUESTIONS INVOLVED RESPECTING THE SO-CALLED " HE-
REDITARY PREDISPOSITIONS TO DISEASE " YET FULLY UNDER-
STOOD AND PROPERLY INTERPRETED ?

A point may now be raised concerning which some misapprehension seems to exist among medical men (judging from
remarks which I occasionally read or hear expressed). I refer
to the relationship of actual squint to nervous disturbances, and
constitutional conditions which are unquestionably associated
with heredity.

No one can deny that people frequently live for long
periods of time in houses impregnated with sewer-gas, and in
the most malarious regions, without apparently suffering in
consequence. Yet no intelligent man would attempt to prove
to-day that sewer-gas poisoning and malarial infection were
delusions simply because some people had escaped their influence.

The argument has been advanced that, because some cross-
eyed people have escaped epilepsy, chorea, insanity, and functional neuroses of the milder types, it is erroneous to maintain
that eye-strain has anything to do with these conditions. This
is absurd upon its face. The hint might, perhaps, be pertinently dropped, in this connection, that cross-eyed people practically suffer but little from their muscular error, simply because
they have *habitual double vision, which no effort on their part
can correct.* These subjects learn very quickly to practically
discard one image (the one seen by the crossed eye) and to use
one eye only for ordinary vision. In other words, they never
try to blend the images of the two eyes, except in certain attitudes of the head which result in a single visual image without
an effort on the part of the patient.

It is only in those cases where (in spite of a muscular
error) the images of the two eyes can be blended by a great
effort that the patient begins to experience the deleterious
physical influences of abnormal muscular tension in the orbit.

If we admit the proposition that eye-defects, or anomalies of the ocular muscles, are liable to become causes of
impaired nervous energy (because they demand an excess of
nervous expenditure), we are forced to the conclusion that, the
earlier this source of physical depression is removed, the better

are the prospects of the person so relieved of escaping diseases which impaired nervous energy necessarily tends to hasten or develop. We are naturally led to question if the so-called "neuropathic predisposition" is not dependent (in a certain proportion of cases, to say the least) upon eye-strain.

We might possibly, also, be led to think that the so-called "*tubercular tendency*" (which is present, as far as my observation goes, in nearly 50 per cent. of all cases of marked functional nervous disease) might, in some cases, be modified, controlled, or perhaps arrested before its physical results become apparent, by taking from the life of such subjects a load which their small reserve-capital of nervous energy particularly unfits them to endure.

It is hard to give up the view, so universally conceded, that a predisposition to disease means a "constitutional taint." Yet, in many cases, we are absolutely unable to demonstrate that any evidence of physical weakness or disease has appeared until sufficient time had elapsed from the date of birth for the development of a serious impairment of nervous energy. What has caused it? Has it been deficient nourishment, a lack of maternal care or solicitude during childhood, gross violations of the rules of hygiene, or a lack of prudence on the part of the individual when of matured experience? The history of case after case answers "no" to such surmises. These, then, are not the all-important factors in every case. Phthisis, epilepsy, chorea, headaches, neuralgias, hysteria, dyspepsia, obstinate constipation, nervous prostration, inebriety, and many other evidences of the neurasthenic state are markedly hereditary. What is the load (if any) which many sufferers of this type are carrying through life? *Have they a congenital burden,*—which is, perhaps, too often unrecognized? I leave these questions for future research to solve.

Did the reader ever see a tired horse fall prostrate under an excessive burden? *How long would he remain so were the burden not removed?*

More people die of phthisis than of any other disease; hence anything that bears upon the tubercular predisposition is of the most vital interest to the community at large. While the possibility of this disease being directly transmitted (*per se*) by heredity has been called into question, and decided probably in the negative, there is, on the other hand, a general concurrence of medical opinion that a *predisposition to "consumption"* is inherited, and that the children of tubercular parents are peculiarly prone to be short-lived.[1]

Again, the *predisposition to gout* is unquestionably transmitted by heredity. Yet, if we are asked to explain in what that predisposition lies, we are forced to avoid the point at issue by such platitudes as "that the processes of assimilation are weak and defective," "that the formation of uric acid is excessive," "that the liver fails to do its duty," "that the kidneys are sluggish," etc. We know, however, that sometimes the poor wretch in a hospital gets as frequent attacks of gout, and often larger chalky deposits, than does the rich wine-drinker; that medication often proves of little or no avail in restraining the disease; that attacks come on in spite of a most rigid diet and regular habits of life; that gouty symptoms do not appear (except in rare cases) before the age of 20, and usually much later in life; and that a *condition of low nervous. vitality* is almost always one which co-exists with a gouty diathesis.

Now, I have lately found in several cases of typical chronic gout, where chalky deposits have been very extensive and a history of direct heredity could be traced through several generations, that an examination of the eyes has revealed a very high degree of eye-strain. In my opinion, such a condition is not exceptional, but rather universal; hence I am led to think that this factor is certainly one not to be despised nor disregarded. If eye-strain exist, it can unquestionably cause a state of low nervous vitality; it is constantly demanding an excessive expenditure of nerve-force by the patient, who, with this particular diathesis, requires all the nerve-power at his command to aid the processes of digestion, assimilation, and excretion; it has been an active and oppressive handicap from the date of the patient's birth, and, if unrelieved, will remain so until he dies; it is a tangible hereditary defect that has doubtless passed, like certain characteristic features peculiar to the family, through some of the patient's ancestral lines; finally, it has probably more to do with the inheritance of digestive ills late in life than any other one factor that has ever been observed in this type of patients.

---

[1] I am at present engaged in collecting clinical statistics upon this point from the inmates of a prominent sanatorium for consumptives.

It should constantly be borne in mind that no two cases exhibit identical manifestations of nervous depression or irritation. Some patients who are suffering from such conditions manifest the effects in physical, others in mental, disturbances. The heart's action may be alone disturbed in some cases, the stomach may give out in others; some may complain alone of spasmodic muscular troubles, some may notice its effects in the eyes, some are rendered sleepless, others may suffer from more or less persistent pains, a few complain alone of skin disturbances, and so on throughout the different parts of the entire human organism.

The reader can understand how these apparently discordant facts may be reconciled when he recalls the fact that by means of the brain and spinal marrow, and the nerves which unite these centres to the different parts of the body, we are enabled to see, hear, taste, smell, appreciate touch, swallow, breathe, and perform voluntary muscular acts. It is by means of our nerves alone that the heart beats; the digestive processes go on without our knowledge or control through the same agencies; the blood-vessels contract and dilate in accordance with the demands for blood telegraphed to the nerve-center by different organs and tissues; and every process pertaining to life is thus automatically regulated. It requires no medical knowledge to see at once how a disturbance of so complicated an electric mechanism as the nerve-fibers and the nerve-cells of a living animal are can upset all or any one of the individual functions enumerated. Many of our houses are furnished to-day with electric bells by means of wires distributed in the walls. In some houses we light the gas-jets, and even the rooms themselves, by means of the same subtile fluid. When the battery becomes weak, or when the wires are disarranged or broken, what may be the results? Some of the bells may cease to ring when the button is touched, while others work properly. Perhaps the electric light may fail in some rooms and burn with its accustomed brilliancy in others. The gas-jets may not be properly ignited. So it is with the nervous apparatus of man. From the same cause one patient may have nervous dyspepsia, another sleeplessness, a third headache or neuralgia, a fourth weakness of the muscles, a fifth disturbances of sensations, a sixth hysteria, chorea, or epilepsy. It is needless to multiply illustrations.

The nervous system of man has been very aptly compared

to a mountainous region where any atmospheric disturbance
calls forth a " series of echoes " at distant points.   So it is with
many of the so-called "functional diseases."   They may be
simply the manifestations of a disturbance of the nervous system, entailed by causes which have been overlooked or imperfectly relieved.

Many diseases, which are to-day commonly regarded as of
bacterial origin, owe their development, in my opinion, to some
underlying cause that has impaired the nervous functions, and
thus rendered the patient peculiarly susceptible to deleterious
atmospheric influences.   This view is held by many others
besides myself.   It is gaining ground among the profession in
England and France, and Dr. Thomas J. Mays, of Philadelphia, has had the temerity to discuss, in two public lectures,
whether phthisis is not to be regarded as a pure neurosis.[1]
Some points in his theory seem to me to be rather untenable ;
but many of his observations respecting the clinical association
of the tuberculous predisposition with various marked neuroses
are unquestionably accurate and in accord with well-accepted
facts.

I have personally given this subject considerable attention,
because in my maternal ancestry phthisis has been extremely
frequent, and the duration of life materially lessened thereby.
I am also confident that a correction of a high degree of latent
hypermetropia (whose existence was unsuspected until atropine
was instilled into my eyes) marked a turning-point in my
own physical state which has never ceased to be a cause for
gratitude.

For many years I have carefully investigated the ocular
conditions of every patient who had cause to fear, as I had done
in the past, the dreaded advent of pulmonary consolidation
and softening.   I have never yet encountered a case of typical
phthisis in which eye-strain did not exist as a factor (more
or less potent, in my opinion, in causing and hastening its development).   I believe most firmly that, had this factor been
recognized early in life, before the eyes were employed in study
and other occupations, and if all anomalies of refraction and
muscular equilibrium had been thoroughly rectified at that
time, many of the hopeless sufferers from phthisis that we all
have met would have escaped the disease ; not because certain

[1] Therapeutic Gazette, November and December, 1888.

atmospheric germs would not still have constantly assailed them, but because the vigor of their constitutions would not have been so sadly impaired by a constant and useless expenditure of nerve-force as to unfit them to combat disease.

The time is surely destined to come when legislation (influenced by public opinion) will step in to protect the young from the serious physical effects of eye-strain. Sooner or later we will see nearly every child subjected to an eye-examination before admittance is sought for and obtained to our higher grades of instruction. Not a mere perfunctory set of tests, but a thorough search for existing defects based upon scientific methods, and made by physicians skilled in that line of work.

In this millennium a myopic child will no longer be placed in the row farthest from the black-board and be held responsible for the stupid mistakes caused by his inability to see. No longer will the excessively "hypermetropic" child struggle along with headaches, blurring of the type on the page, mental confusion and distress after a prolonged use of the eyes, and the thousand other ills it is compelled now to endure from the ignorance of its parents or medical adviser. No longer will "esophoria," "exophoria," and "hyperphoria" [1] be unrecognized or deemed as of trivial importance; nor will an actual "squint" (of far less clinical importance, although a deformity) be magnified into undue importance.

Life-insurance examiners will then deem it necessary to take into consideration the possible existence of these hidden factors of disease before they pass final judgment; physicians will, in time, rely less on drugs as specifics and study symptoms more intelligently from the physiological stand-point.

A very pertinent remark was lately made by one of our most polished orators, when he said: "Many people go through this world like those who ride backward in a railroad-train. They never see anything until they have passed by it." So it is with many in our profession. Great advances are always made in spite of bitter opposition, groundless prejudice, and willful misconstruction. Those who stood highest in professional esteem at the time incited the populace to rage when Jenner first advocated vaccination. Harvey was despised because he

---

[1] The reader is referred to a subsequent page (Chapter II) for an explanation of these terms.

advocated the circulation of the blood. McDowell was condemned to professional ostracism when he first practiced ovariotomy, yet to-day a costly monument, erected by the medical profession, marks his resting-place.

In closing this chapter I would aim to make the following deductions clear to the reader:—

1. Eye-strain arises chiefly from defects in the refraction of the eye and an imperfect equilibrium in the muscles which move the eyes.

2. These conditions, when present, tend to cause an excessive expenditure of nerve-force by the individual in direct proportion to the amount of defect to be overcome.

3. Excessive expenditure of nerve-force upon any one organ is commonly made at the expense of some other organ, or, if not, is paid out of the "reserve" amount of nerve-capital possessed by the individual.

4. The extent of the drafts thus made upon the "reserve-" capital and the amount of "reserve-capital" are the two factors which can alone determine, in any individual case, how long this state of things can last without causing a "nervous bankruptcy."

5. The conditions mentioned as those which chiefly tend to cause eye-strain are transmitted from parent to child; hence they become operative at birth and last until death, unless mechanically or otherwise relieved.

6. They are capable of detection and accurate measurement during life by scientific procedures. Some of the methods employed by oculists in testing the eye-muscles are not worthy of perpetuation.

7. A condition of exhausted nervous vitality is sure to impair the general health in many ways, and to render the individual more liable to disease than when in full vigor.

8. Many of the constitutional diseases which ultimately imperil the lives of their victims are indirectly the result of a state of low nervous vitality (a state which is frequently the result of eye-strain, from well-understood causes that might have been easily recognized and relieved).

9. The so-called "inherited predisposition" to certain diseases is unquestionably based, in many cases, upon some anomaly of the visual apparatus. I am so well convinced of

this fact that I assert it (without fear of contradiction) from carefully gathered statistics.

10. The examination of the eye for errors of refraction and accommodation, and a thorough familiarity with the tests lately advocated for the detection of anomalies of the ocular muscles ought not to be confined exclusively to the practice of the oculist.

They are as valuable to the general practitioner as are the physical signs of the chest.

CHAPTER II.

THE TESTS OF VISION AND OCULAR MOVEMENTS.

THE steps which should be employed in examining the eye for errors in refraction and accommodation, as well as those employed to ascertain and measure defects in the power of ocular muscles, have not thus far been discussed.

I expect to offer nothing new. What I shall say may possibly seem to the specialist elementary and incomplete.

It is impossible in a book of this kind, which is designed rather for general practitioners and beginners in ocular work than for specialists, to describe with full details how to employ the more elaborate instruments, such as the ophthalmoscope or ophthalmometer; or to go into all the tests for refraction and anomalies of adjustment of the ocular muscles with exhaustive completeness. I hope, however, to make the details of an ordinary eye-examination simple and within the comprehension of all.

The importance of this department of diagnosis can hardly be overestimated in nervous maladies. It has been my custom for many years to examine the vision of nearly every patient sent to me; as my experience has shown me many times that remarkable cures may be made by the light thus shed upon the causation of obscure nervous symptoms.

Unfortunately for the sick, in many instances, physicians in general seem to think that the examination of the eye is too difficult a field for them to intrude upon without months of special preparation for it. While this is undoubtedly true, in case the ophthalmoscope and some other instruments of precision are to be employed, it is by no means a difficult matter for a person acquainted with physics to acquire a practical and satisfactory knowledge of the tests here described in a comparatively short time and with but a limited number of patients, provided that he works faithfully and intelligently. The healthy (?) as well as the sick can often be used to familiarize the beginner with the practical adjustment of prismatic, spherical, and cylindrical glasses, and also with the tests employed to detect *heterophoria* (defective equilibrium of the eye-muscles).

(22)

Defective vision does not always produce ill-health; hence among friends or in the immediate family the beginner may find a field for investigation and practice.

TESTS FOR REFRACTION.

In the first place it is not absolutely necessary to have a complete Nachet case of lenses. Such a case is expensive. By

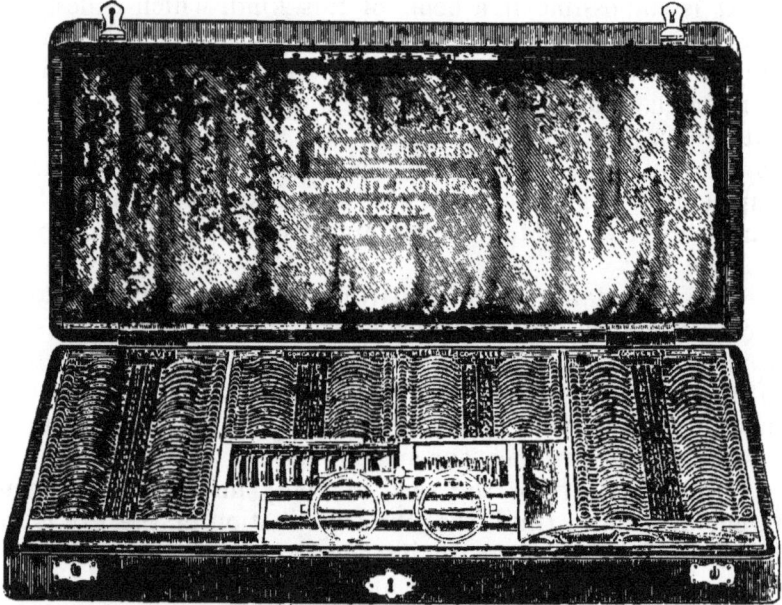

FIG. 3.—NACHET'S COMPLETE SERIES OF TRIAL-LENSES, MOUNTED IN SILVERED AND GILT RIMS.

Contents: 30 pairs of convex spherical lenses, from 0.25 to 20.0 D.; 30 pairs of concave spherical lenses, from 0.25 to 20.0 D.; 18 pairs of convex cylindrical lenses, from 0.25 to 6.0 D.; 18 pairs of concave cylindrical lenses, from 0.25 to 6.0 D.; 10 prisms, 2 to 20 degrees; 4 plain colored glasses; 1 white glass; 1 half-ground glass; 2 metal discs, with stenopaic slit; 1 stenopaic disc, with hole; 1 solid metal disc; 1 improved adjustable trial-frame, with revolving cells and graduated scales; 1 double-grooved graduated frame. This case should have an improved adjustable trial-frame in place of the Nachet frame, as it has the more modern mechanical adjustments.

selecting a limited assortment of lenses and prisms, different combinations can be made to meet the needs of almost every eye-defect encountered in medical practice.

There is furnished, with the various small cases designed by prominent oculists, a sheet of Snellen's test-types for distance, and also one containing several paragraphs printed in an assortment of types of various sizes, to be used as a test for reading-power. Each paragraph is numbered; so that a record can be kept of the one read by the patient as a test. These test-type slips can be purchased separately, however, of any optician. It is best to have each *mounted on card-board ;* and it is well to have the one used in testing for distance a double one, with different letters on the opposed sides. If you suspect that the patient is *using his memory* rather than his sight during the tests employed, the board can then be exposed upon different sides at various periods of the examination.

Z

S A

P X E

K F O L

R T B V E

L P H S H 4

G F E D O E 3

You will find that the letters are mathematically made for testing distant vision. Above each line a numeral or Roman character is placed to designate the *number of feet or meters at which the normal eye should read the line with ease.* Thus, the large letter on the top line will be designated usually by CC or 60, while small letters of the lower line will be marked X or 3. This shows that the top letters should be read easily at two hundred feet, or sixty meters, by the normal eye, and the lower line at ten feet, or three meters. After the physician has provided himself with a good trial-case, a set of prisms, and the necessary test-type, let us see how he should proceed with an examination of a patient's vision.

FIG. 4.—SNELLEN'S TEST-TYPE.

Greatly reduced in size to admit of being used as an illustration.

TESTS FOR SUSPECTED HYPERMETROPIA OR MYOPIA.—First, hang upon the wall the test-type for distance, and place the patient with his eyes on the same level and at a distance from it of exactly twenty feet. Then take the double-grooved or triple-grooved spectacle-frame from the trial-case and insert a plate of metal in the left rim of the frame, so that when it is used by the patient the left eye will be covered. Then place this frame upon the patient's nose and ask him or her to read aloud the letters on the testing-sheet from the top downward, line by line. This act tests the vision in the right eye. Note (while the patient reads) the following facts: (1) if *all the letters are read properly ;* (2) if the patient *reads without apparent effort ;* (3) at *what line* the patient *fails* to read.

Now make a record as follows: O. D. (oculus dexter, or right eye) $V. = \frac{20 \text{ (feet)}}{- \text{ (type)}}$. The dash in the fraction is filled with the number which indicates the last line which the patient reads. When the vision is normal, the fraction will be as follows: $V. = \frac{20}{20}$ or $\frac{20}{xx}$. If the patient fail at the line next above the normal point, the fraction would be expressed by $\frac{20}{30}$ or $\frac{20}{xxx}$.

Remember that the *numerator* represents the distance, commonly in feet, but occasionally in meters, between the patient and the test-type, and that the *denominator* represents the numeral on the test-card placed above the last line of type read by the patient (which indicates the normal distance in feet or meters at which it should be legible to the normal eye).

Now, if the vision of the right eye is found to be defective, try and improve it and, if possible, to render it normal, or as nearly so as possible, by testing the effects of concave or convex glasses upon it as the case seems to indicate, beginning with the weakest lenses and gradually increasing their strength until the vision reaches its highest acuteness. This takes some little practice and experience. If *convex glasses* are found to be indicated, note the *strongest* which gives the best vision to the patient; if *concave,* record the *weakest* glass that overcomes the defect.

When perfect vision cannot be obtained by spherical lenses, astigmatism may be suspected. It is well, in such cases, to add convex or concave cylinders to the best spherical already found, until normal vision or an approach thereto is obtained. Each cylinder must be revolved in front of the spherical until that axis is found which gives the patient the most distinct vision. The accurate determination of astigmatism will be discussed later.

When the vision is found to be normal ($\frac{20}{20}$ or even $\frac{20}{15}$) without any glass it indicates that there is no myopia; but there may still be a slight degree of astigmatism or possibly a high degree of hypermetropia. In this case it is wise to try convex lenses until the strongest glass is found with which the patient can read the test-letters as well as he could without a glass.

Many persons normally have vision equal to $\frac{20}{15}$; and I make it a rule never to stop the examination of refraction when

$\frac{2.0}{2.0}$ is obtained if any glass can be found which will give $\frac{2.0}{1.5}$ vision to a patient.

In some cases the reader may find himself unable to obtain normal vision in either eye of a patient by means of cylindrical or spherical glasses. I presuppose a certain degree of acquired facility on his part with glasses of the forms specified, and a carefully made effort to overcome the existing defect.

In such a case it is well for a beginner or a general medical practitioner to consult some expert oculist (if near at hand), and thus to ascertain the results of an *ophthalmoscopic examination*. The patient may have some mechanical impediment to vision, such as an opaque lens within the eye (cataract) or an opacity of the cornea; or he may have a high degree of astigmatism, which can often be estimated with some accuracy by the ophthalmoscope, retinoscope, or ophthalmometer. Again, he may be found to be suffering from morbid changes within the optic nerve or the retina.

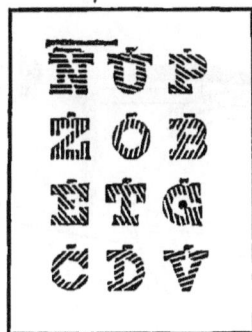

FIG. 5.—ASTIGMATIC LETTERS.

When it is found that a patient is so blind in an eye as to be unable to recognize any of the letters on the testing-card at any distance, it is well to note (before sending him to an oculist) if he can recognize with accuracy the *number of fingers* which you hold before the eye and to record the results of such investigation. This test should be made with the fingers in all possible positions in reference to the diseased eye (directly in front, above, below, and to either side of it).

The results of an examination of a suppositious case, up to this point, might be recorded as follows:—

O. D. V. $= \frac{20}{\text{xxx}}$ (manifest) made $\frac{20}{\text{xx}}$ by $+ 1.25$ D.[1]

The word "*manifest*" in this record means that the far-sightedness, or hypermetropia, which *apparently* exists, is overcome by a convex ($+$) glass which focuses at thirty inches ($+ 1.25$ D.). After the use of atropine any increase over this

---

[1] The letter D in this record stands for the term *diopter*, corresponding to about a glass of forty-inch focus. The numerals that precede the letter D indicate that the glass employed was one and one-fourth diopters in strength, or about thirty-inch focus.

amount which may be developed is recorded as "*latent*" far-sightedness.

The reader will understand, when I exhibit the method of recording such observations more fully, why it is that the right and left eyes have to be separately examined and corrected (as already described) before the binocular vision is tested with and without the needed correction. I usually make, upon the page of my own case-book, a note, prior to the use of atropine, somewhat as follows, in a suppositious case :—

O. D. (right eye) $V. = \frac{20}{xxx}$ (manifest) made $\frac{20}{xx}$ by $+ 1.25$ D.

O. S. (left eye) $V. = \frac{20}{xl}$ made $\frac{20}{xx}$ by $- 1.25$ D.

Binocular $V. = \frac{20}{xxx}$ made $\frac{20}{xx}$ by *this combination.*

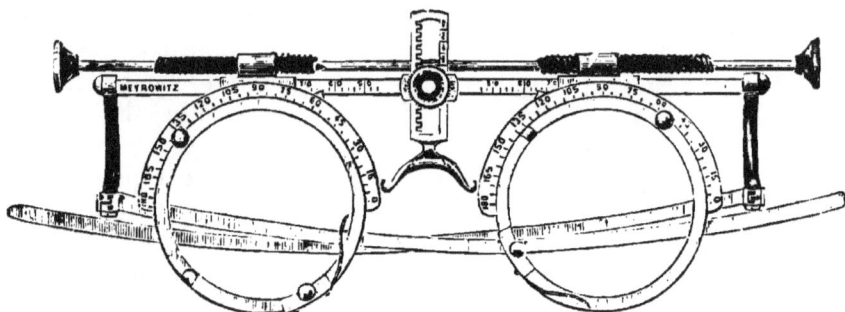

FIG. 6.—IMPROVED TRIAL-FRAME, WITH REVOLVING CELLS AND GRADUATED SCALE.

The pupillary adjustment is effected by a right and left triple-thread screw motion, and the distance between pupils can be read on the sliding-bar above the cells; the nasal adjustment is made by means of a rack and pinion, the height of nose being indicated on the guide of the nose-piece; the temples are attached to the frame in such a manner that the front can be swung to any angle and the lenses brought close to the eyes.

Such a record of a suppositious case would show that the patient was apparently *far-sighted*, or "*hypermetropic*," in *the right eye*, and *near-sighted*, or "*myopic*," in *the left eye*. It would lead me to believe, also, that the right eye (when under the influence of atropine) might show a still greater defect, which is now rendered "latent," or hidden, by an excessive development of the muscle of accommodation, and to suspect that the apparent myopia in the left eye might, possibly, be somewhat decreased or even absent when the effects of atropine were established.

In all far-sighted eyes nature tries, from the date of birth,

to compensate for the congenital defect (an eye which is too
flat; see Fig. 1) by an hypertrophy or *enlargement of the ciliary
muscle* (see Fig. 2); hence, when this muscle is temporarily
paralyzed by atropine, the true refractive condition of the eye
is no longer masked by ciliary spasm. Far-sighted patients,
therefore. lose their clearness of vision more or less at once
when atropine is used. The normal or the " myopic " eye, on
the contrary, is but little affected (as regards the outline of
objects seen at twenty or more feet from the eye) by the use of
atropine, although excessive light may annoy the eye in any case.

TESTS FOR SUSPECTED ASTIGMATISM.—An irregular curvature
of the cornea or lens is a common defect of the eye known as
astigmatism. In a large percentage of cases the astigmatism
is of the corneal type; hence the special instruments devised for
its detection are based upon reflections from the surface of the
cornea that vary from those of the normally-constructed eye.

Because, in the astigmatic eye. the surface of the cornea
or lens is not that of a section of a perfect sphere, the images
of external objects thrown upon the retina are distorted in
proportion to the degree of astigmatism.

Of late years much has been written concerning the effects
of astigmatism upon not only the use of the eyes, but also upon
the bodily health. So far has the pendulum of opinion swung
in this direction that many oculists of repute are to-day correcting by a glass one-fourth to one-half a diopter of astigmatism.

While the author doubts personally whether very low
grades of astigmatism (when uncomplicated by heterophoria)
should necessarily entail upon a patient the wearing of a glass,
still, he is willing to admit that he has seen many cases where a
very weak astigmatic glass has afforded great comfort and relief.

If, during the examination of an eye. the vision of a patient
cannot be brought to the normal point by any spherical glass,
there is a justifiable suspicion that the case may be one of astigmatism. Under such circumstances it is advisable to hang up
some form of astigmatic test-card. at a distance of twenty feet
from the patient's eye, and to give this card bright and equally
distributed illumination by sunlight or artificial light.

The most common card in use is the so-called clock-face
card (Fig. 7). although a card of astigmatic letters (Fig. 5)
and various other devices may be employed in testing astigmatic
patients.

If the patient is astigmatic to a marked degree, some of the lines on the clock-face will appear distinctly black, others less black, and some possibly indistinct and of a light-gray color.

The card containing the two black lines (c, Fig. 7) should now be placed over the dial with the lines directly over the

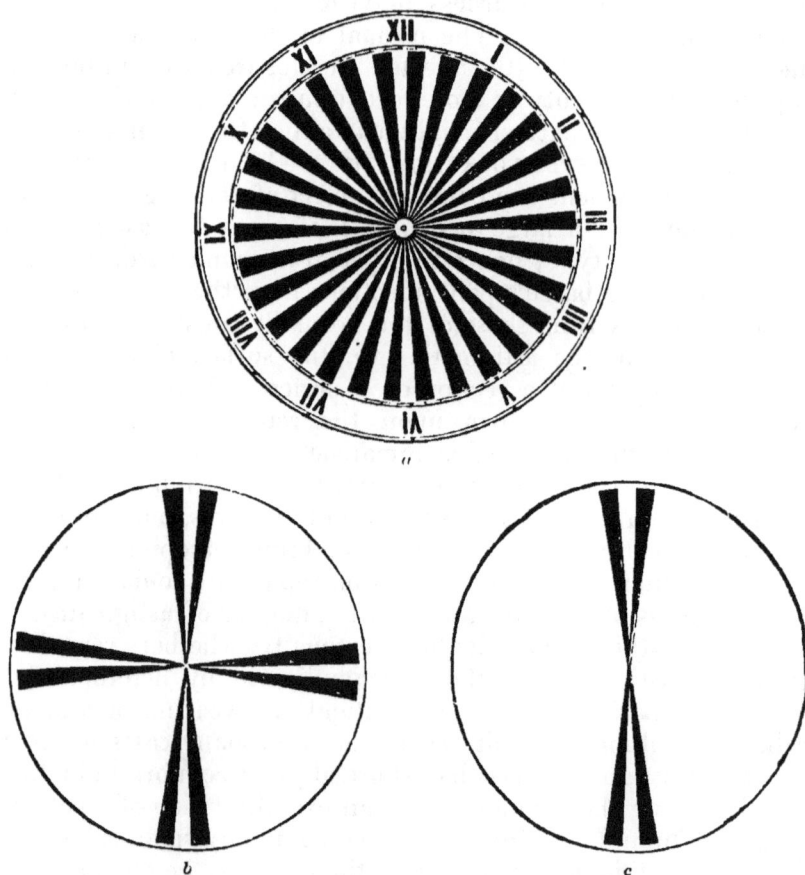

FIG. 7.—NOYES'S SELECTION OF GREEN'S SET OF ASTIGMATIC TESTS.
Consisting of a clock-dial mounted on heavy card-board and three circular discs (a, b, c) which can be attached to it.

lines of the clock-face that look blackest to the patient. Spherical lenses are placed over the eye of the patient until these lines are perfectly distinct and *clear-cut down to the centre.* The card is then rotated at right angles to the former position and again spherical glasses are applied until the lines become distinct in this position. The difference in strength between the lenses that give the best vision in each meridian will determine the strength of the cylinder that corrects the astigmatism, and with this cylinder all the lines on the original clock-face should be equally distinct.

FIG. 8.—PLACIDO'S DISC, OR KERATOSCOPE; DISC MADE OF ALUMINUM.

This instrument provides for a rapid method for the detection of suspected astigmatism. Its circles are reflected in the cornea by holding the disc at a distance of ten inches from the eye of the patient. Distortion of the circles in the reflection indicates astigmatism.

A card with lines at right angles (b, Fig. 7) can be used, which renders the rotation of the card unnecessary; but it is usually better to fix the vision solely in one meridian at a time by using the card with only one set of lines. The cylindrical glass thus found, added to the spherical glass that gives the best vision, will usually raise the vision to the normal ($\frac{20}{20}$ or $\frac{20}{15}$).

A very careful and accurate estimation of hypermetropic or myopic astigmatism by means of a trial-case and test-cards only should always be attempted after the pupil has been widely dilated with a solution of homatropine or atropine. Such examinations are far more precise than when the pupil is of normal size, because it is easier to get the exact axis of the cylindrical glass through a large opening than a small one, as well as the exact amount of myopia and hypermetropia that exists in addition to the astigmatism.

The author has seen many cases where the results of the Javal ophthalmometer have been materially modified by the use of atropine. If the astigmatic defect is in the lens of the patient, the instrument devised by Javal fails to measure it.

The instrument known as Placido's disc (Fig. 8) is sometimes useful in the hands of a novice for detecting high grades of corneal astigmatism; but the instrument of Javal is by far the best of all that have yet been devised for rapid and accurate work.

The instrument of Javal consists of a large metal disc, intricately divided, with figures upon it reversed. Through the center of this passes a barrel with an adjustable eye-piece, through which the observer looks. In front of it stands an adjustable head-rest for the patient, with an arrangement of lights for a strong illumination of the disc. In front of the disc and attached to the barrel of the instrument is a curved metal bar carrying two adjustable " mires " that slide on the bar. Attached to this bar also is a " pointer " (painted white) that extends to the circumference of the metal disc, and thus indicates the axis at which it points. The bar with its mires and pointer can be revolved by the operator throughout the entire circumference of the disc.

When this instrument is used, the patient is placed with his head in the head-rest with one eye looking directly into the barrel of the instrument, and the other eye covered with a blackened metal shield. A bright illumination of the disc, mires, and pointer is insured by the lights attached to the head-rest.

The operator now looks through the barrel of the instrument and regulates the disc and eye-piece so as to see two distinct reflections of the illuminated disc and mires overlapping each other on the anterior surface of the patient's cornea, as in a mirror.

The figures on the disc are reversed by reflection from the cornea and are now read with ease. The movable mire is slid along the bar until its inner edge joins that of the fixed mire, with the pointer at zero. If the middle dividing-line of both mires is not now continuous, the mires are rotated until the lines are continuous. The point at which the pointer then rests indicates the *meridian of greatest or least curvature.*

The mires are then turned to a point exactly 90 degrees from this point, and the mires, if there is astigmatism present, will be overlapped or separated. If the mires overlap, the astigma-

FIG. 9.—JAVAL-SCHIOTZ OPHTHALMOMETER (AMERICAN MODEL) WITH ELECTRIC ILLUMINATION.

tism is "*with the rule,*" and each step on the fixed mire that is overlapped by the movable mire represents one diopter of astigmatism. If the mires are found to be separated they must be again approximated and the mires rotated back to the original position, when they will again lap and the amount of astigmatism "*against the rule*" can be read. If the *astigmatism is* "*irregular*" there will be a distortion of the mires and lines on the disc that makes it impossible to get the cross-lines on the mires continuous in the opposite meridians.

It is advisable, in the opinion of the author (after the use of Javal's instrument), to *confirm the diagnosis* made by it by placing the proper lenses over either eye of the patient (after the pupil has been dilated) and testing the vision of each eye separately by means of the various astigmatic cards and also by Snellen's test-types (Figs. 4 and 5).

It is usually advisable to *correct all astigmatism found* by means of *cylindrical glasses for constant wear*, combining with the cylinders, if necessary, such spherical glasses for the correction of existing myopia or hypermetropia as seem best adapted for constant wear by the patient.

Let us now suppose that, during the first examination of a patient, we have examined each eye separately, carefully corrected all existing errors found, and succeeded in getting $\frac{20}{xx}$, or normal vision, for each eye separately; that we have then tried both eyes together with the glasses best adapted for each, and found the patient able to read the normal type for distance without fatigue or conscious effort; and, finally, that we have made a careful record of each point noted during our observations. Are we now prepared to order glasses for the patient? Have we noted all that is important to note? To both of these inquiries I would say to the beginner, emphatically, "No." Several steps still remain to be taken, even before the use of atropine (which it is generally best to employ before a final decision is arrived at).

TESTS FOR HETEROPHORIA.

Until within a comparatively few years the necessity of carefully measuring the power of adduction, abduction, and sursumduction of the eyes, and of determining the presence or absence of muscular anomalies in the orbit, particularly in "nervous" subjects, seems to have been practically disregarded even by oculists. Even to-day this form of defect (which probably is, as a rule, congenital) seems to be omitted from prominent mention among the enumerated list of etiological factors of nervous symptoms by some authors of note. In some cases I have known it to be overlooked even by ophthalmologists of world-wide reputation, simply on account of a careless and hasty examination for the defect. It is an extremely common defect of the eye, and may prove a very serious one to the patient. It is an important factor in many subjects afflicted with headache; it often exists, to a high degree, in epileptics; it is frequently found among children who suffer from chorea; it may unfit a patient for sewing, reading, attending places of amusement, or using the eyes in any way for any length of time. I have known it to cause vomiting and so-called period-ical "bilious attacks" by exciting a reflex irritability of the central nervous system. One patient of mine (a close student) was completely cured of chronic dyspepsia by the use of prisms, which corrected an insufficiency of six degrees of the external recti muscles; he now uses his eyes without fatigue, and all bodily ailments have disappeared without the use of drugs.

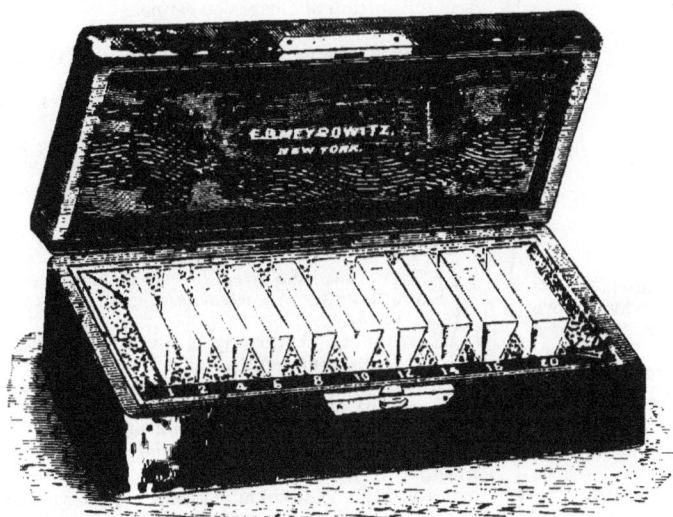

FIG. 10.—COLLECTION OF TEN SQUARE PRISMS.
1, 2, 4, 6, 8, 10, 12, 14, 16, and 20 degrees.   Size, one and one-half inches.

In order to properly investigate the condition of the ocular muscles, several tests have to be made. I do not personally regard *any of these alone as sufficient for diagnostic purposes.* The tests which I advise you to invariably employ are as follow :—

1. MOVEMENT OF EYES.—Direct the patient to look fixedly, with both eyes, at some small object (say, the point of a pencil), *and to follow it as it is moved* before the face of the patient, at a distance of about ten inches. Watch both eyes carefully at the same time, and note if a *tremulous movement* in either eye is

present in any position of the eye as it moves about, and if the two eyes act in perfect unison with each other.

The following table may be incorporated here with possible benefit to the reader, in following the description of the various tests employed to determine errors of refraction or of muscular adjustment in the orbit.

Terms related to the focus of the eye (*refractive terms*).	HYPERMETROPIA (*far-sightedness*). A *shallow* eye (from the front to the back), causing an imperfect focus of objects. MYOPIA (*near-sightedness*). An *elongated* eye (from the front to the back), causing an imperfect focus of objects. ASTIGMATISM. An *irregularly curved cornea*, or *lens*, causing distortion of images on retina. EMMETROPIA. A *perfectly constructed eye*.
Terms related to the muscles which move the eyes (*muscular terms*). These were first suggested by Dr. George T. Stevens.	ESOPHORIA. A tendency of one or both eyes to deviate toward the nose. EXOPHORIA. A tendency of one or both eyes to deviate toward the temple. HYPERPHORIA. A tendency of one eye to rise above the level of its fellow. ANAPHORIA. A tendency of both eyes to assume too high a plane. KATAPHORIA. A tendency of both eyes to assume too low a plane.
	Strabismus.[1] { ANATROPIA. An *actual turning* of both eyes upward. KATATROPIA. An *actual turning* of both eyes downward. ESOTROPIA. A turning of one or both eyes inward (*convergent squint*). EXOTROPIA. A turning of one or both eyes outward (*divergent squint*). HYPERTROPIA. An actual rolling of one eye upward or downward (*vertical squint*). }
	HETEROPHORIA. Abnormal adjustment of the eye-muscles. ORTHOPHORIA. Normal adjustment of the eye-muscles. ADDUCTION. The power of the internal muscles of the eyeballs. *It varies in health between 25° and 60°.* ABDUCTION. The power of the external muscles of the eyeballs. *It should be 8° in health.* SURSUMDUCTION. The power of the vertical muscles of the eyeballs. *The right and left should be alike.*
Various forms of glasses employed by oculists.	SPHERICAL. Ground upon a *convex or concave sphere*. Used to correct hypermetropia and myopia. CYLINDRICAL. Ground upon a *convex or concave cylinder*. Used to correct astigmatism. PRISMATIC. *Two plane surfaces of glass meeting at an angle.* The thick side is termed the base of the prism. Used to relieve errors of adjustment of the eye-muscles.

---

[1] The chief point of clinical discrimination between the *strabismic* and *non-strabismic cases* is this: In strabismic cases, when a plain, red glass is placed over one eye, *unconquerable diplopia exists.* It is often difficult to make strabismic patients find the second visual image, because they tend to suppress the image of one eye: but, with patience and the use of the red glass, double candle-flames can be disclosed to this type of patients, and their relative positions made clear and positive.

2. EXCLUSION TEST.—The patient is now instructed to *fixedly gaze* with both eyes at an object twenty feet from the eye, while the physician *shields one eye* with a card or sheet of paper so as to exclude the object from view. He observes at the same time any deflection or trembling of the covered eye. If deflection or trembling occur, it indicates a weak muscle.

When the shield is moved quickly so as to alternately cover the field of the right and left eyes, it is important to watch the covered eye behind the shield, and to notice where it rests when covered, and if it tend to move inward, outward, upward, or downward at the moment when the shield is shifted to cover the opposite eye. This is known as the "exclusion test"; and a *marked jump* of either eye or both is an important clinical observation.

Whenever an eye shows a decided "jump" on shifting the card over the opposite eye, it indicates that the eyes are not in equilibrium; that, as soon as one of the retinal impressions is removed by covering either eye, the covered eye assumes a position in the orbit that differs from the fixed eye; and the "jump" observed on shifting the shield is due to an effort on the part of nature to again regain the retinal impression of the test-object.

A jump inward (at the moment the card is shifted) indicates that the eye, when covered, tends to diverge from its fellow (exophoria); a jump outward indicates a marked tendency toward a convergence (esophoria); a jump in the vertical direction points strongly toward a marked difference in level between the eyes (hyperphoria); provided that the eyes jump in opposite directions. If they both jump in the same direction it indicates anaphoria or kataphoria.

3. THE RED-GLASS TEST.—If a piece of plain red glass be held in front of one eye of some patients and they are instructed to look naturally with both eyes open at a candle-flame twenty feet from the patient, the candle-flame may appear as two distinct flames,—a white and a red one! In other words, such a patient will show an *unconquerable diplopia* (double vision) *when the red glass is employed*, of which he is probably ignorant, and which does not exist when the red glass is withdrawn from before the eye.

Again, if the red glass fail to cause diplopia when the eyes are directed at an image directly in front of the patient, it is always well to instruct the patient to keep his eyes fixed on the test-flame, and *to turn his head strongly both to the right and left* while the red glass is kept between one eye and the candle-flame. In this way diplopia may be detected at the extreme right or left position of the head, and the variety observed (homonymous or crossed diplopia) should be recorded.

A diplopia (with simply a red glass before one eye) *constitutes the dividing line between strabismus and heterophoria.* Constant diplopia only exists in strabismus.

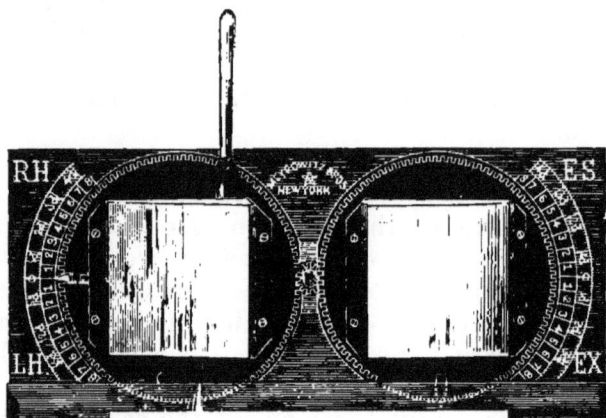

FIG. 11.—STEVENS'S IMPROVED ROTATING PRISM-SLIDE (PHOROMETER).

The slide contains two cells, in each of which rotates a disc, each disc carrying a prism of 5°. Each disc is furnished with a border of teeth or cogs. A small gear-wheel, placed between the two discs, communicates the movements from one disc to the other. Around the outer part of the border of each cell is a narrow raised band, on which is marked a scale of degrees, increasing from the center each way from 0 to 8°, the numbers representing the refracting *angle* of prism,—the method of notation now commonly used. A second scale, just outside the first, is similarly graduated according to the new system of designating prisms, by the refracting *power*, or number of degrees of minimum deviation. This scale is made to correspond to the standard recommended by Drs. Jackson, Dennett, and others to the American Ophthalmological Society at its session in 1888. The numbers in this scale are marked 1° D., 2° D., etc. The scales represent a far greater degree of accuracy and uniformity than the prisms found in the trial-cases in common use, and the convenience of the two scales is apparent.

Directions for Using the Slide with Rotating Prisms.—Place the slide in the groove of the phorometer with the side on which the scales of degrees are seen *from* the patient. The side marked *RH.* and *LH.* will then be before the patient's right eye, and that marked ES. and EX. before the left, the instrument being brought to perfect level. To determine *hyperphoria* bring the lever of the prism-slide to the vertical position. The pointer will then be at 0. The patient, looking through the two glasses at an object (a lighted candle placed at a distance of twenty feet) directly in front, sees double images of the object. Should one of the images appear higher than the other the prisms are caused to rotate until the images are brought to the same horizontal plane. By making the rotation slowly it will, in many instances, be carried farther than when the correction is quickly made. The pointer then indicates the form and degree of manifest hyperphoria. To examine for esophoria and exophoria bring the lever to a horizontal position and then make adjustment until the images are in an exact vertical line.

4. TESTS FOR ESOPHORIA AND EXOPHORIA.—After the preceding tests, I would suggest that the reader place upon the
patient a spectacle-frame previously arranged with a disc of
ordinary glass tinted red to cover one eye, and a prism of 5°,
with its base directed vertically downward, before the other eye.
Then direct the patient's vision upon a candle-flame at a distance of at least twenty feet. The prism causes *two candles to
appear* (one being colored red by the glass of that hue), both
of which to the normal eye should be seen *in a vertical line*.
If the red image is seen to that side of an imaginary vertical

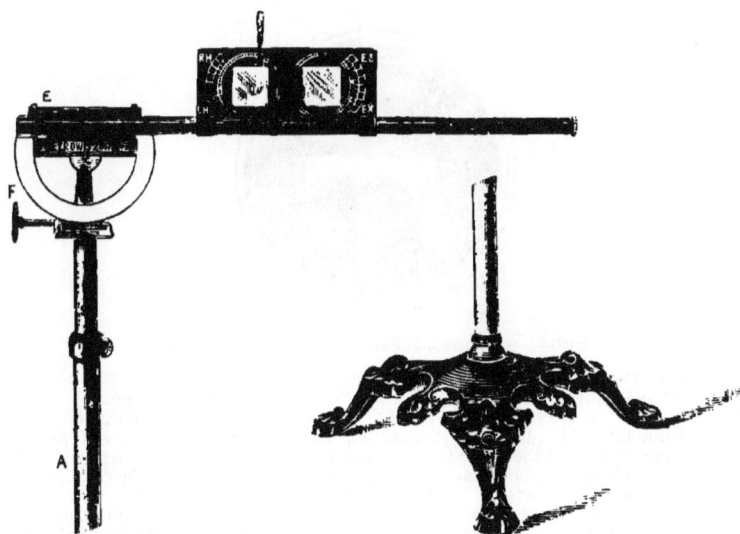

FIG. 12.—SAME INSTRUMENT AS FIG. 11, WITH IMPROVED STAND.

line dropped through the white image which corresponds to the
eye with the red glass, the external recti are insufficient; if the
red image is seen on the opposed side of the same imaginary
vertical line, the internal recti muscles are weak.

Any deviation of the candle which exists can be remedied
easily by placing a prism with its *base outward* before one eye
for insufficiency of the externi (esophoria), and with its *base
inward* for insufficiency of the interni (exophoria). I usually
note both the weakest and the strongest prism which corrects
the candle-deflection.

To those who can afford to purchase a "*phorometer*" for

the purpose of measuring anomalies of the ocular muscles, greater accuracy and the saving of much time will be assured. The tests (2, 3, and 4) are but poor substitutes for a phorometer, but they can be employed by those who have but a trial-case of lenses and prisms.

5. TEST FOR HYPERPHORIA.—Prior to the general employment of the phorometer, considerable difficulty was experienced in making accurate observations upon a patient who was suspected of having the condition known as hyperphoria.

In order to be sure that the two eyes do not tend to assume the same plane, it is of vital importance that the prisms placed

FIG. 13.—MADDOX'S TEST FOR HETEROPHORIA.

It consists of a hard-rubber disc mounted in a metal rim of the size of trial-lenses. so as to fit easily into the trial-frame. which holds in the center a glass rod. The effect of this transparent cylinder is to cause an apparent elongation of a single flame into a thin line of light, quite dissimilar from the flame itself, as seen at the same time with the other eye ; so that there remains practically no desire to unite the two images, whose relative position thus easily indicates the condition of equilibrium of the two eyes. The line is always at right angles to the axis of the rod ; so that to produce a vertical line. with which to test horizontal deviations. the rod is placed horizontally, and to produce a horizontal line. to test vertical deviations, it is placed vertically. The test is made prettier, and any desire for single vision still further reduced, by placing a red glass before the other eye.

before the patient's eyes to cause homonymous images should be absolutely horizontal. In order to insure this, various devices were early employed, among which may be mentioned the fixation of the head of the patient by a photographer's rest, and the attachment of a spirit-level to a specially-constructed frame designed to hold the prisms.

These obstacles to scientific precision of measurement have been entirely overcome by the employment of the phorometer.

A simple device known as the *Maddox cylinder-rod* (Fig. 13) has also proven of value in my hands as a substitute for the phorometer. It is far more accurate than the use of prisms without the phorometer in the determination of hyperphoria and other forms of heterophoria.

The preceding cut illustrates this addition to a trial-case of lenses. It consists of a metal disc containing a slit three-fourths inch long by one-eighth inch wide. Over this slit is placed a

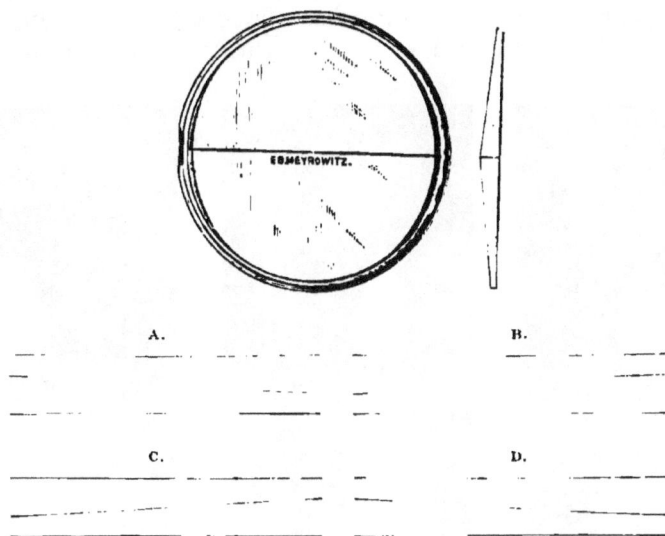

A.

B.

C.

D.

FIG. 14.—SAVAGE'S MODIFICATION OF MADDOX'S DOUBLE PRISM.

This instrument consists of two six-degree prisms fitted base to base into a trial-lens rim. It is specially adapted to the detection of a want of equilibrium of the oblique muscles. The prism is placed before one eye with the axis vertical, the other eye being covered, and the patient is asked to look at a horizontal line eighteen inches distant. The effect of the double prism is to make the line appear to be two, each parallel with the other. The other eye is now uncovered, and a third line is seen between the other two, with which it should be perfectly parallel.

The eye before which no prism is held is considered as the one under test. With the double prism before the right eye the patient is asked about the position and direction of the middle line. It may be nearer the bottom, thus showing left hyperphoria; or, again, it may extend farther to the right than the other two and not so far to the left, thus showing exophoria, or *vice versa*, showing esophoria. If the right ends of the middle and bottom lines converge, while the left ends diverge, the superior oblique of the left eye is at once shown to be in a state of underaction. A represents such a test of the left eye; B shows a test of the left eye when the inferior oblique is the too-weak muscle; C represents a test of the right eye, the loss of parallelism between the lines being due to underaction of its superior oblique; D, the same condition of the inferior oblique of the right eye.

cylinder of glass which causes the retinal image of a candle-flame to assume the appearance of a brilliant band of light.

When this cylinder-disc is placed over one eye of a patient with the rod vertical, the uncovered eye sees the candle-flame and the covered eye a brilliant band of light running horizontally. If no hyperphoria exist, the band of light should run directly through the candle-flame, and a one-degree prism, with its base up or down, placed before either eye, should cause the candle and bar of light to assume different planes.

On the other hand, if hyperphoria exist, the bar of light will appear either above or below the candle (the higher eye seeing the lower image). A prism with the base up or down,

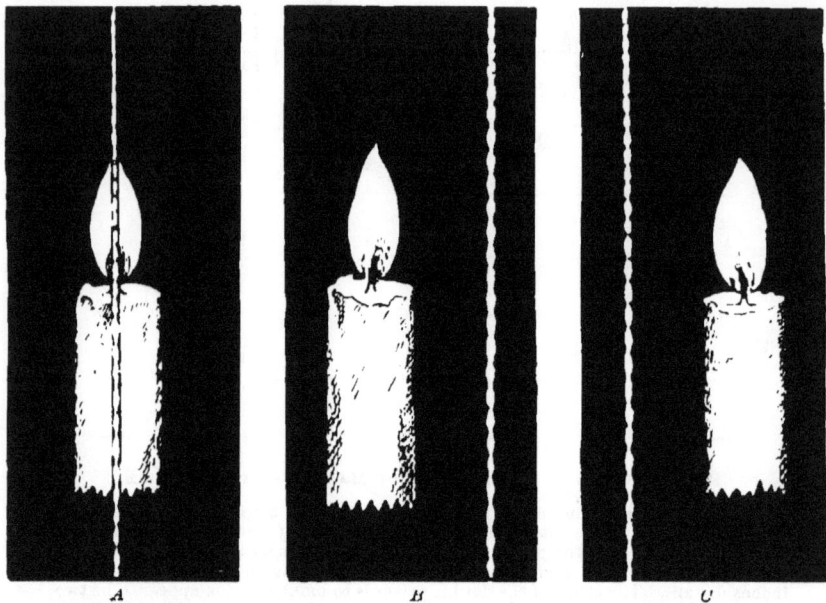

FIG. 15.—MADDOX'S ROD-TEST FOR HORIZONTAL DEVIATION. (De Schweinitz.)[1]

The rod is before the right eye. *A*, the line passes through the flame,—orthophoria. *B*, the line passes to the right of the flame,—latent convergence, or esophoria. *C*, the line passes to the left of the flame,—latent divergence, or exophoria.

of sufficient angle to bring the images to the same plane (the band cutting the flame at the center) will then denote the degree of " *manifest* " hyperphoria.

By means of the Maddox cylinder-rod, when placed horizontally before one eye, " *manifest* " esophoria or *exophoria* can also be approximately determined. The band of light in this

[1] Figs. 15 and 16 from de Schweinitz on "Diseases of the Eye" (W. B. Saunders, Philadelphia). By permission of the author.

case takes the place of a prismatic deflection of an image described in tests 3 and 4 (pages 36 to 39).

Clinical experience seems to show that an accurate determination of hyperphoria in any given case is of vital importance; hence it is imperative that a beginner does not come too hastily to a conclusion respecting heterophoria. It is not uncommon to observe a tendency to both a vertical and lateral deflection in the same patient,—conditions which are termed "*hypereso-phoria*" and "*hyperexophoria.*" When it comes to the treatment of such a case, either by prismatic glasses or graduated tenotomy, it is usually wise to investigate and, if possible, to correct the hyperphoria first.

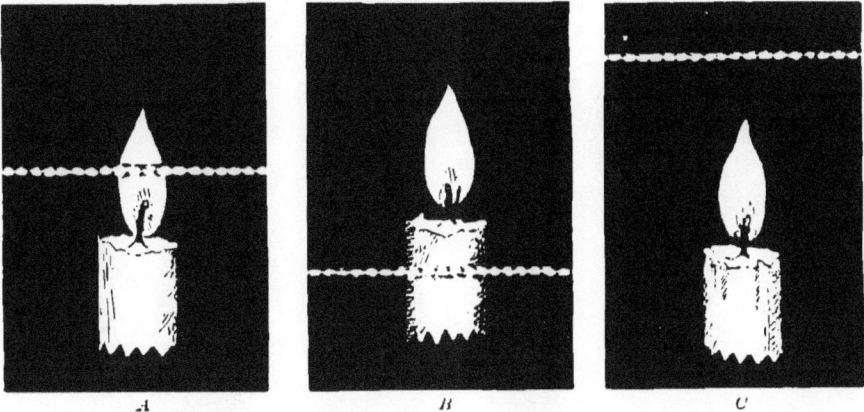

FIG. 16.—MADDOX'S ROD-TEST FOR VERTICAL DEVIATION. (De Schweinitz.)

The rod is before the right eye. *A*, the line passes through the flame,—orthophoria. *B*, the line passes below the flame. The upper image belongs to the left eye,—right hyperphoria. *C*, the line passes above the flame. The upper image belongs to the right eye,—left hyperphoria.

I would advise any reader who meditates devoting his attention to the determination and treatment of heterophoria to provide himself with a phorometer. I am satisfied that accurate and scientific work in this line cannot be done without the aid of this valuable instrument.

6. TESTS FOR ADDUCTION, ABDUCTION, AND SURSUMDUCTION.—Before we pass to the consideration of the next step which should be employed in the first examination of a patient's eyes,—*i.e.,* the *determination of the adduction, abduction, and sursumduction,*—it may be well to hastily summarize the tests which have already been described and presumably employed.

Up to this point the reader is supposed to have obtained information respecting the manifest refractive conditions of each eye (see page 35) and to have placed before each eye of the patient, in the trial-frame, such glasses as are accepted by the patient. These manifest errors of refraction may, or may not, be correct,—a point that must subsequently be determined by the use of atropine.

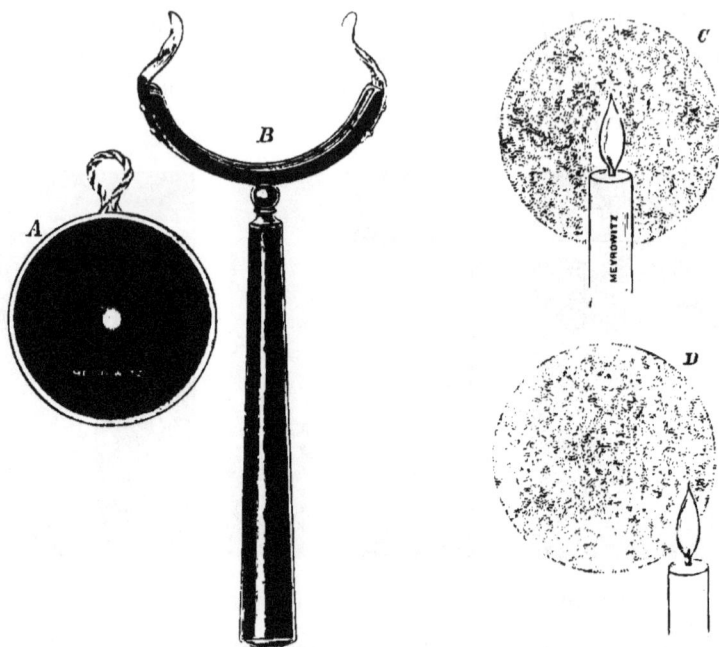

FIG. 17.—*A*, STEVENS'S LENS FOR THE DETERMINATION OF HETEROPHORIA, WITHOUT HANDLE. *B*, HANDLE. *C*, ORTHOPHORIA. *D*, HETEROPHORIA.

The inventor's description of this instrument is as follows: "In the determination of the various tendencies of the ocular muscles it is often advisable, and even necessary, to bring to our aid as many forms of evidence as can be made subservient to our purpose. While the phorometer remains pre-eminently the reliable and efficient working instrument in the determination of heterophoria, auxiliary means are often required to confirm or to explain its indications. We sometimes also require an instrument for making provisional examinations more portable than the phorometer. As such an auxiliary and provisional instrument I have devised the stenopaic lens, which possesses manifest advantages (*A*). The purpose is to present contrasting images to the two eyes. With the lens the image of a candle-flame twenty feet distant, seen through the stenopaic opening, is a large and perfectly-defined disc of diffused light. If, for the purpose of effecting a diffusion, we employ the uncovered convex lens, a very slight movement of the lens, in or out, up or down, gives to it the effect of a prism in those various directions.

"If a convex lens, about 13 D., is covered, except at the optical center, where a circular opening of three millimeters or less diameter acts as a stenopaic window, the small opening serves the double purpose of preventing an adjustment of the lens as a prism and of cutting off the hole in such a manner as to give the impression of an exact disc of light bordered by a frame. A metal or hard-rubber disc of the size of the lens of the trial-case, perforated by an opening of the required diameter, and supplied with a perfectly-centered lens, is a convenient form. It may be used with a handle (*B*), enabling the patient to hold it in his own hand, or it can be placed in the trial-frame.

"In orthophoria the untransformed image should be found exactly in the center of the disc. In heterophoria it will tend toward or beyond the border. If the flame sink below or rise above the center, while at the same time it deviates laterally, we thereby discover, by a single comprehensive view, all the elements of a compound deviating tendency, so far, at least, as that tendency is manifest (*C* and *D*). In this important respect the stenopaic lens presents a feature both unique and of much significance. While, by other methods of inducing diplopia or contrast, we may discover, at a distance of some meters, first one and then the other element of a deviating tendency, by this instrument all the collective elements are presented simultaneously to the eye, thus eliminating a very important source of error. In respect that it is simple, cheap, and small enough to be carried in the vest-pocket, and that it, more than any of its class, represents the true relation of the visual lines, it is a useful test. Its disadvantages are those common to every instrument held close to the eye when in use in these examinations."

The presence or absence of tremulous movement in either eye and the results of the exclusion-test have next been observed and recorded. Following this an examination has been made for the different varieties of heterophoria (by means of the phorometer and Maddox rod-test) that are disclosed by the patient on the first examination. The information gained, up to this point, regarding the heterophoria may, or may not, be correct; but it is important to record it as a basis for comparison with subsequent examinations.

We have now reached a point where the power of the interni, externi, and vertical muscles of the orbit should be very carefully measured and recorded. Much valuable information is often afforded by this step.

For the purpose of measuring the power of ocular muscles, *square prisms* (Fig. 10) possess great advantages over round ones. They are much more easily handled, and there is a greater certainty of maintaining the proper axis of the prisms to the eye, especially when several prisms are being simultaneously handled.

The prisms which I employ in my office are one and one-half inches square, of uniform sizes, and ground on the edges. My case includes ten prisms of the following degrees: 1°, 2°, 3°, 4°, 5°, 6°, 8°. 10°, 15°, and 20°.

I sometimes employ a stereoscopic box, whose pupillary distance can be regulated by a screw. The patient can hold this before his eyes while looking through it at the test-object (a candle-flame at twenty feet from the eye). Into this box prisms can be dropped or withdrawn at will, with the base of the prism in any direction. This box is of use chiefly when testing patients with high adducting-power.

In measuring the abduction or sursumduction square prisms are readily held in the hand before the patient's eye.[1]

In order to test the ADDUCTION, place before the eyes of the patient a prism, with the base toward the temple, of sufficient angle to cause double images to appear of a candle-flame twenty feet from the eye. Then instruct the patient to draw the two images together. If he fail to do so, decrease the strength of the prism until he can accomplish it. Should he succeed, increase the strength of the prism gradually until he is unable to hold a single image.

By directing the vision of the patient to an object (such as your finger) within two or three feet of the eye, until that object becomes single, and then suddenly withdrawing it, the patient is often aided in learning to exert his interni upon a distant object.

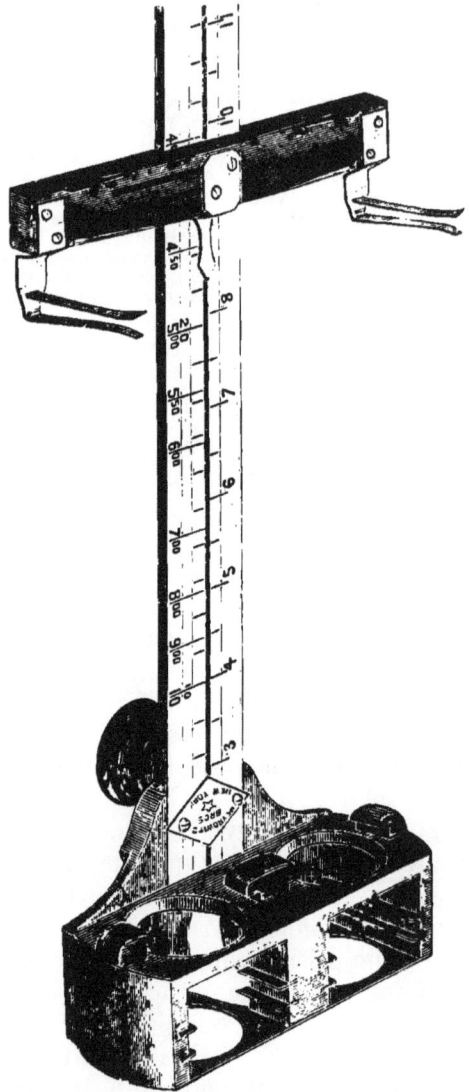

FIG. 18.—NOYES'S PRISM-HOLDER.

[1] The *round prisms of a trial-case* are very apt to lead the examiner into error, if used for this purpose. Their axis is apt to be changed by getting rotated in the frame, and they are awkward to handle, when several are used on top of each other in building up prisms before the eyes of a patient.

The *normal adduction varies from 25° to 70° of prism*, according to practice and education. When it falls below 15°, and all efforts to increase it seem futile, there is reason to suspect either an exophoria, a complicating hyperphoria, or a latent hypermetropia.

To test the ABDUCTION. prisms should be placed before the eyes of the patient, with their bases toward the nose, as long as the patient is able to hold a single image.

There seems to be more uniformity in the normal power of the externi than of any other set of ocular muscles. *The normal power of abduction seldom exceeds a prism of 8°.* Any great excess over that amount is liable to cause unconquerable crossed diplopia. A *low abduction* almost invariably indicates that *some* "*latent*" *esophoria exists*, even in those patients that, at first, fail to disclose any manifest heterophoria. This statement is quite clearly demonstrated in the published records of some of the epileptic cases.

The SURSUMDUCTION in any given case is obtained by placing prisms, with their bases down, before either eye of the patient, until the limit of the power of maintaining single images is reached. The sursumduction is recorded as "right" or "left," according to the eye tested. *Each should be alike ; and neither should, as a rule, exceed a 4° prism*, except in highly myopic patients.

7. TESTS OF VISUAL ACCOMMODATION.—After the refraction has been carefully determined at the distant point, and before atropine is used, the *vision should be tested at the reading-point.*

For this purpose, a series of test-types has been arranged by Jaeger, which vary in size from very small type (No. 1) up to very large type (No. 13).

The full refractive correction for distance, having been put in a trial-frame, is placed upon the patient. A shield is then placed over one eye and he is asked to read the finest type (No. 1). If he can do so the type should be moved steadily toward his eye until the closest point at which he can read it distinctly is found and recorded. Then it should be removed from the eye until the farthest point of distinct vision is obtained. Before the age of 40, when accommodation begins to fail, the patient should read the No. 1 type at from six to twenty inches.

If the patient cannot read the No. 1 type readily, convex spherical glasses should be *added to the distance correction* until the glass is found which gives the best range of accommodation with the greatest comfort to the patient. The same strength glass added to the other eye will usually be found to give equally good vision. If this is not the case, there is reason to suspect that there is more latent hypermetropia in one eye than in the other, which can be determined when the eye is under atropine. This test may be recorded, in a suppositious case, as follows: Reads Jaeger's No. 1, at eight to twenty inches, with + 1.00 over distance glasses.

The accommodation begins to fail, to a perceptible degree, at 40 years of age, and is usually completely lost at 55 to 65

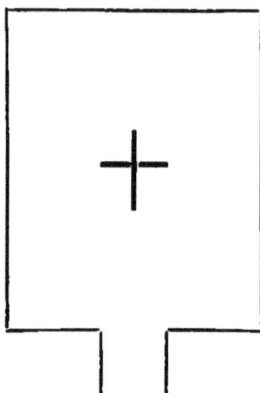

FIG. 19.

years of age; hence convex glasses of gradually increasing strength are usually required for reading after the 45th year. This loss of accommodation is usually in exact proportion to the age of a patient; so that, the age being known, the glass can usually be selected at once. At 45 a patient usually requires + 1.00^s, and at 50 years of age + 2.00^s, for reading, when no refractive error exists in either eye.

The failure of accommodation from age is called "*presbyopia.*"

8. TESTS FOR HETEROPHORIA IN ACCOMMODATION. — After the reading-glass has been carefully determined the muscular tests for heterophoria should be made at the reading-point with the correcting glasses on. These are called "*tests in accommo-*

dation," and are made on the same principle as the tests for esophoria and exophoria, already described, except that, in place of the candle at twenty feet, the line-and-dot test of von Graefe at fourteen inches is commonly used.

A more convenient method is the metal plate with a cross upon it, which is attached to the phorometer (Fig. 19).

The description of the various tests, already mentioned, will enable the reader to fill out the following blank. This is similar to one used by me to record the results of examinations of the eyes of my own cases.

. . . ., *189* .

Mr. . . . .

Residence . . . .

$V.=\begin{cases} R. \\ L. \end{cases}$

Esophoria . . . . *In accommod.* . . . . . .

Exophoria . . . . *In accommod.* . . . .

Abduct. . . . . *Adduct.* . . . .

Hyperphoria, R., ... *L.,* . . . .   .

Sursumduct., R., ... *L.,* . . . .

With red glass . . . .

. . . .

Glasses ordered for distance :

         O. D. . . . .

         O. S. . . . .

Glasses ordered for reading :

         O. D. . . . .

         O. S. . . . .

Ophthalmoscopic examination . . . .

. . . .

Now, when we have carefully examined our patient respecting all the data indicated in the preceding table, are we safe in passing an opinion respecting the condition of the eyes? I would again say "No."

9. ATROPINE TESTS.—We have now reached a point where we should *administer atropine* to the patient. I usually employ a solution of gr. iv of sulphate of atropine to an ounce of distilled water. This can be kept constantly in the office in a

phial with a rubber top dropper substituted in place of a cork. A drop or two in each eye will suffice, in most subjects, to dilate the pupil widely and to paralyze the power of accommodation of vision for near objects in about three hours.   In occasional instances it becomes necessary to keep the patient under its influence for several days, but this is not the rule.

In cases where it is inadvisable or extremely inconvenient for the patient to use atropine, *homatropine* may be employed. In at least 75 per cent. of my cases I have found it fully as efficient and satisfactory as atropine for determining latent errors of refraction.   As the effect of homatropine is practically gone in twenty-four hours. the advantage to the patient is considerable.   I usually employ a 3-per-cent. solution of hydrobromate of homatropine.   This should be dropped into the eye three or four times at intervals of fifteen minutes. in such a way as to allow it to flow over the cornea.   Half an hour after the last instillation the loss of accommodation is usually complete.

When sufficient latent errors of refraction are disclosed, after the employment of this drug. to justify a doubt regarding the full extent of the refractive error thus found, a solution of atropine may. if deemed wise. be instilled into the eyes of the patient later, in order to eliminate any possible question regarding the refraction.

It is well to caution the patient, after using atropine, that he may possibly suffer from the sunlight, and that colored glasses will relieve him of this annoyance.   It is also best to tell him that his vision may become very blurred for distant objects, in case he is far-sighted ; and that, in any case, he will be *unable to read or to write by the aid of vision without glasses* for several days.   I have known hypermetropic patients to become greatly alarmed at the rapid loss of vision which has followed the use of atropine, all of which could easily have been avoided had they been prepared for it by timely words of explanation.   It is always well to explain to far-sighted subjects the difference between " manifest" and " latent" hypermetropia. and to make them intelligent as regards the effect of atropine upon the " focusing" muscle before you administer it.   If they are forced by their business to use their eyes for near work, while under the influence of atropine. a pair of cheap glasses may be given them for temporary use while under its influence.

I cannot impress too strongly upon the reader the necessity

4

of using atropine upon a patient (if young) for diagnostic purposes when an error of refraction or of accommodation is suspected. Personally I *do not regard an examination as complete without it.* It solves the question of the presence of latent hypermetropia,—a very common defect and a very serious one (from the stand-point of the neurologist), if allowed to go unrecognized. It reveals the existence of a previous ciliary spasm. It often arrests headache as if by a magic touch, and solves the nervous origin of many other similar symptoms.

Patients who boast of their acuteness of vision and who, apparently, justify their statement by reading test-type at a distance, without the aid of glasses, are often astonished and sometimes alarmed at the immediate loss of this power, which is brought about by the use of atropine. This surprise is heightened when (by the use of proper lenses) their power of vision for distance is immediately restored, and they become conscious, for the first time, of the muscular effort which they have been compelled, in the past, to exert in order to see without them. I shall never forget, personally, the sensation which I experienced of " seeing without effort " when a latent hypermetropia was discovered in my own eye and corrected by glasses.

These experiences are well-known facts among oculists, but to the profession at large they often occasion as much of a surprise as to the patient.

I could point to case after case, in my own experience, where the cause of neuralgic attacks, excruciating headache, vomiting, extreme nervousness, and many other symptoms (not, apparently, connected with eye-defect) would have remained unrecognized if atropine had not been employed. There is a rule among most oculists,—viz., to give to a hypermetropic patient the *strongest convex glass* with which he can comfortably read the normal test-type (xx) at a distance of twenty feet. It is impossible, in many cases, to decide this fact without atropine or an ophthalmoscope.[1] The former method is the best one,

---

[1] There are two sources of error which are possible in all ophthalmoscopic examinations as a step toward the determination of refraction. The first of these is that the observer may not be able to perfectly relax his own accommodation while using the instrument. Most oculists of large experience believe that they can do this with certainty,—a belief which is, perhaps, not always well founded. The second source of danger lies in the accommodation of the patient. This cannot always be relaxed by instructing the patient to look at an object twenty or more feet distant from the eye. I am satisfied that mistakes in the determination of refractive errors by the ophthalmoscope are far more frequent than are generally supposed. For the past twelve years I have examined the eyes of almost every patient intrusted to my care by the aid of the test-type, after the pupils have

because the accommodation of the oculist, as well as that of the patient, has to be excluded in the latter, and it has the advantage that it can be employed by the general practitioner as well as by the specialist.

It is well known and generally acknowledged, to-day, that the *ophthalmoscope is not an instrument of precision when the refraction of an eye is to be positively determined.* The greatest ophthalmoscopist of his day, in this country, tried, some years ago, to determine the refraction of the writer's own eyes by this instrument, and made a dismal failure, as he himself had to confess after atropine was employed.

Now, after the patient returns with widely-dilated pupils, the physician should *carefully repeat each step of the previous examination.* He should record the results of these tests and then compare them with those obtained before atropine was employed. If the eye is a normal one the vision will be $\frac{20}{xx}$ after atropine has been used, as it was on the first examination; but, when an error of refraction or accommodation exists, changes of a greater or less degree may be noted. He may find, moreover, that the power of adduction and of abduction of the eye will be modified in some patients by the action of the drug upon the accommodation of vision, and that a different degree of heterophoria may be detected. He can now decide intelligently as to the glass which is best adapted to restore vision for distant and near objects in each eye of the patient, and he is prepared to advise the patient respecting the use of the glasses selected.[1] He can decide, also, respecting the question of the utility of prisms or of tenotomy, if the patient has marked manifest heterophoria. You can judge more accu-

---

been fully dilated by atropine. I am not aware that I have ever lost the confidence of a patient by the use of this drug. In my experience intelligent persons are always willing to submit to a temporary inconvenience for the purpose of obtaining *positive information* respecting any point that is deemed of scientific value in relation to themselves. I have personally come to regard the ophthalmoscope as an unreliable instrument for the determination of refraction. Its use is rendered compulsory, however, in very young children and in those who, from ignorance or feeble-mindedness, are unreliable in their reading of test-type. It is generally accepted, furthermore, among our best oculists, that astigmatism (a recognized source of nervous perplexity) is always estimated more accurately with the pupil widely dilated by atropine than with the normal pupil. The reasons which I have already given must suffice to explain why the use of atropine constitutes a most important preliminary step to the detection and estimation of any error in the eye-muscles, although many other arguments might be brought forward to prove its advisability in some subjects.

[1] The advisability of a full correction by glasses of existing hypermetropia can only be decided after the condition of the patient, his age, his susceptibility to reflex irritation from eye-strain, etc., have been carefully considered. It is not usually advisable to force a young subject to wear a glass which fully corrects the *latent* hypermetropia.

rately respecting the proper angle of the prism required, in case
their use is indicated. I would caution you, however, against
deciding this latter point *before the error of refraction* (if such
exist) *is corrected.* I have seen patients who gave evidence
of marked heterophoria (5° to 8°), when the refractive error
was uncorrected, exhibit no such defect when glasses which corrected that error were worn. Prisms, in such a case, would
inflict injury upon the patient rather than afford relief.

I would remark that views which are here advanced respecting the dependence of abnormal nervous phenomena upon
eye-defect are not new. They are in antagonism, however, to
those of some authors, and have been more or less actively
combated of late, especially in regard to eye-defect as a cause
of chorea and epilepsy. I do not think the relationship between
eye-strain and attacks of headache or neuralgia can be denied,
although it is only hinted at by Anstie, and is omitted by most
authors who have written on the causes and cure of these distressing maladies. Some of our best neurologists, as well as
most oculists, are now investigating with renewed interest not
only the ametropic conditions of the eye, but also the eye with
heterophoria.

Facts are being daily substantiated beyond dispute which
met with ridicule some years since. Every day, in my own
experience, I am strongly impressed with the curative effects
of glasses in various forms of functional nervous disturbances. In my opinion, the neurologist of to-day who fails to
familiarize himself thoroughly with the examination of the eye
omits an evident line of duty both to himself and his patients.
No neurologist can send all of his cases to an oculist for an
opinion, and, even if he could do so, he should, at least, be able
to verify the opinion thus gained respecting the *refractive errors*
found and the *state of the eye-muscles.* He requires a case of
lenses and prisms in his office as much as an electrical outfit,
and he should know how to use both,—the one as an aid in
diagnosis and both often as a means of cure. Personally I have
come to regard the examination of any patient sent to me as
incomplete until I have tested the state of refraction and accommodation, and examined with care the condition of the ocular
muscles. This view has not been hastily formed, and my daily
experiences confirm me in it. I believe the time will come when
the tests employed in eye-examinations will rank in importance

in neurology with the knee-jerk test, which for generations, as Gowers remarks, simply "amused school-boys."

10. THE INVESTIGATION OF LATENT HETEROPHORIA BY PRISMS.—To obtain a clear insight into the form of heterophoria (p. 35) that exists in any given case requires repeated and careful examinations. To determine the exact extent of the heterophoria is even more difficult, because many patients disclose at first but a small part of the muscular anomalies that exist.

The heterophoria that is rapidly disclosed during the first examination, or before prismatic glasses are worn by the patient, is usually designated as the "*manifest*" *heterophoria*; that which remains undisclosed until the patient is aided by prismatic glasses to abandon his unconscious habits of adjustment (entailed upon him by the muscular anomalies of his orbit) is termed "*latent*" *heterophoria*.

It is with latent heterophoria that we have to deal, if satisfactory results are to be obtained in cases of intractable nervous diseases,—such, for example, as epilepsy, chorea, insanity, insomnia, chronic headache, and extreme neuralgias.

Whenever a patient presents himself for the first time at my office, considerable care is always exercised in studying the posture of the head, the expression of the eyes, in watching for tremulous movement or motion of the eyes in any direction, in studying the position of either eye when shielded behind a card (exclusion test), and in observing the effect of a red glass placed before either eye of the patient.

These tests have been fully described in preceding pages. They often afford very valuable information respecting the heterophoria that exists.

Again, after the refractive errors have been fully determined under atropine, the proper glasses placed before each eye, and the adduction, abduction, and sursumduction taken under such conditions as are most favorable to the patient, it is well to study the power of the individual eye-muscles as a further guide to latent heterophoria.

I always regard a *low abduction* (less than 5°) as peculiarly significant of a latent esophoria. A *marked difference in the sursumduction* points toward a latent hyperphoria. A very *high abduction* (more than 10°) with a rather low adduction justifies a suspicion of latent exophoria.

Patients who show an esophoria at distance and an exophoria in accommodation, or *vice versâ*, are always to be suspected of concealing a *latent hyperphoria*.

Too much care cannot be exercised in patients of this class, prior to the performance of graduated tenotomies for the correction of the manifest heterophoria at distance.

Prismatic glasses are not only inadequate as satisfactory remedial agents in most cases, but they may be positively injurious to certain classes of patients. Strict limitations upon their field of usefulness (not generally taught) seem to be rendered probable by late investigations.

A careful study of the different movements of the eyeball, and of the combination of muscles required to produce some of these must impress even the most casual reader with the idea that an agent (such, for example, as a strong prism) which tends to restrict the movements of any one muscle may do harm if persistently worn. Some patients are peculiarly susceptible to such influences.

I have encountered a large number of patients whose eyes refused to tolerate a prismatic glass. Their symptoms were at once made worse whenever they attempted to correct an existing muscular anomaly by wearing a prismatic glass. On the other hand, many patients are benefited at once by the use of prisms, and suffer no inconvenience of any kind from them. What are we to infer from this statement? Are we to surmise that the prisms were either injudiciously selected or improperly placed, simply because the patient could not tolerate them? I think not. Such might possibly be the case in the hands of a novice, but presumably it is not the case in the experience of one skilled in eye examinations. My own experience in many instances has shown me that a tenotomy of the muscle exhibiting the greatest tension has been followed by a complete cessation of the nervous symptoms for which the patient sought relief, in spite of the fact that prisms prescribed to correct the same error have proved intolerable to the patient, and have markedly aggravated the symptoms.

As the development of a latent heterophoria often constitutes the basis of successful treatment, it is vitally important that the information gained shall be positive and beyond cavil. I deem it important, therefore, that the reader shall grasp the principle upon which prismatic glasses are employed as aids in diagnosis and the necessity of having, as part of the office equipment, a large stock of prisms of various strengths, with their bases lateral and vertical, as well as an assortment of interchangeable frames, into which these prisms can be inserted or removed, according as the investigator deems wise.

As many patients under observation wear spectacles for various refractive conditions, many of the frames used in my office are furnished with short hooks instead of bows, so that they can be easily attached in front of the spectacle-frame. The prisms and frames for my office-work are of uniform size; but the frames should have a variety of nose-bridges and should vary more or less in their pupillary distances.

As these frames are often loaned to patients for a period varying from a few days to a few weeks while complex problems in heterophoria are being investigated, and as a larger assortment is necessary to fit each individual face, the number required by a specialist in this work is naturally great. It is my custom to have them provided with extralong screws so as to facilitate the changing of prisms in the frames from time to time without losing the screws.

In investigating a difficult problem of latent heterophoria by the aid of prismatic glasses, too much stress cannot be laid upon the statement that *under no circumstances should a prism be given to a patient which fully corrects the heterophoria disclosed by that patient.*

Prismatic glasses should be employed simply as aids in diagnosis; and the strength of the prism should be increased from time to time only to an extent that partially corrects the heterophoria that is disclosed.

Let me cite an example: A patient, who is wearing a 1° esophorial prism, discloses on tests 3° of esophoria with an abduction of 5°. It would now be safe to increase the esophorial prism to 2°, because the patient would still disclose 1° in excess of the prism and his abduction is 3° below the normal standard. Should the patient after wearing a 2° prism for a short time disclose 4° of esophoria with an abduction of 4°, the prism might safely be increased to 3°. Thus, from day to day, according to circumstances, should we follow up by prismatic aid a heterophoria of the patient as it is disclosed.

There is another point that should restrict us in too rapid use of prisms,—*i.e.*, the inconvenience which some patients experience in walking and estimating distance when prisms are worn. This rapidly subsides in most subjects; but it is wise to caution a patient when prisms are first worn to be careful in going downstairs, or stepping over a curb, or in looking downward while walking for the first few days that they are worn.

Whenever latent hyperphoria is suspected and prisms are being used as an aid in disclosing it, it should be remembered that each degree of hyperphoria is of much greater clinical importance than the same amount of esophoria and exophoria, and that a *half-degree surplus* in excess of the prism would justify its use in most cases.

In all cases of hyperphoria the difference in the sursumduction of the right and left eye should be double the amount of the prism worn, in order to justify its use for any length of time. For example: a patient wearing a $1°$ prism for right hyperphoria should disclose, when tests are made with his prism on, at least $1\frac{1}{2}°$ of right hyperphoria and the right sursumduction should be at least $2°$ more than the left.

Another point of value may be here suggestive,—*i.e.*, that all tests made by the phorometer upon patients that are wearing prisms should be made with the prisms on rather than when they are removed, the strength of the prism being allowed for when the heterophoria is recorded, and whenever the adduction, abduction, and sursumduction are measured.

It is the custom of many oculists, I find, to remove a prism from a patient before tests are made for heterophoria. This is undoubtedly a mistake. It forces the patient to suddenly and unconsciously attempt to adjust for part or all of his heterophoria. Sometimes this may produce serious nervous symptoms on the part of a patient. I recall a case of a patient who had gradually become accustomed to the use of a $4°$ hyperphorial prism, and who showed a surplus of nearly $2°$ over that prism. He had worn the prism for some weeks with great comfort, and was on the eve of a graduated tenotomy for its relief. At the earnest request of his betrothed he was persuaded to dispense with his prism while attending a social gathering. He rapidly became ill, and within two hours fell unconscious, causing great alarm to his family and friends.

I have frequently known an epileptic, who had been free from convulsive seizures for a long time, to have a severe fit on the removal of a prism upon which he had been resting with comfort.

From what has been already stated, it must be evident to the mind of any candid reader that the method here advanced of developing latent heterophoria does not justify the criticism which has been heretofore too frequently made by oculists of repute, that patients are " forced to adjust for prisms that are not indicated, and that heterophoria is created rather than relieved " by the methods here advocated.

There is a practical and important field for prismatic glasses. They may be loaned from time to time to patients for the purpose either of verifying a diagnosis or of developing a latent muscular error which the physician may be led (by repeated examinations of the patient) to suspect. When they are well tolerated the physician may often learn a great deal by their protracted influence. When they are not well borne, it is advisable, as a rule, to discontinue their use at once. It is often wise to prescribe a prismatic glass, also, for a class of patients who are unable (for one reason or another) to submit, at the time, to tenotomy.

Sooner or later I find that such patients usually return. As a rule, they do so for one of the following reasons: (1) because they have developed an additional latent muscular error, which the prisms naturally failed to correct; (2) because they do not tolerate them well, and are made decidedly worse by their use; (3) because they prefer a tenotomy to the inconvenience of a glass which has to be constantly worn; and (4) because they suffer from eye-fatigue on account of the disturbance to co-ordinate movements of the eyeball. There is no doubt that very many persons with nervous diseases are materially helped (if not radically cured) by the aid of prismatic glasses; but the question naturally arises to my mind, in this connection, " Would they not have been more rapidly benefited and permanently relieved, with far less inconvenience to the patient, by tenotomy ? "

The view is held that *a graduated or complete tenotomy is the only means of permanently relieving abnormal tension of a muscle in the orbit.* There are only two ways of overcoming an abnormal tendency of the visual axes to deviate from parallelism whenever the eyes are directed upon an object more than twenty feet off. One of these is by the aid of a prism ; the other

A. L. RANNEY, M.D.,
315 Madison Ave.,
New York City.

NAME OF PATIENT......

Residence

Referred to me by

EPILEPTIC ATTACKS

Total Severe Medium Light

Date Medication Operations

Nervous Symptoms

PRISMATIC EXERCISE

Adduct Abduct Sursumduction Before Exercise After Exercise
R. L. R. L.

DEFECT IN OCULAR MUSCLES

REFRACTION R. L.

is by tenotomy of the muscle which directly aids in producing and perpetuating the deviating tendency. Whenever prisms are prescribed they afford relief practically in the same way as a "rubber muscle" does in orthopædic surgery; in other words, they compel the muscle which is opposed to the base of the prism worn by the patient to overcome the antagonistic muscle, and also to so adjust the eye as to compensate for the refractive effect of the prism. They practically act, therefore, as a "pulley-weight,"—a mechanical device seen in all gymnasiums. Now, if the wearing of prisms had no deleterious action upon those particular muscles, which, in each case, are not at all at fault, and if they invariably exerted only beneficial effects, this principle of treatment could be more generally applied with benefit. Even then the existence of latent heterophoria might, unfortunately, remain unrecognized for a greater or less period of time, possibly to the serious detriment of the patient. On the other hand, if it is satisfactorily demonstrated that tenotomy has been rendered a safe and accurate method of correcting muscular anomalies in the orbit, a fact has certainly been noted that opens a new and shorter route to relief. Such a step enables us, moreover, to decide the question of latent muscular defects in any given case.

It may be an aid to the reader to keep on hand some blanks such as are used by me in my office for

recording my cases.   It allows the tests to be recorded for a
number of days on the same sheet, which can then be filed.   It
was designed by me for neurological cases only ; but it is capable
of modification by specialists in other lines.   I employ occasionally certain pasters over the epileptic headings when I wish to
record the dates and character of some other forms of nervous
attacks.

Bilateral Deviations of the Eyes in the Vertical Meridian.

Within the past two years the attention of oculists has
been drawn by my friend, Dr. George T. Stevens, of New
York, to certain abnormal conditions of the orbit in which
*both eyes simultaneously tend to assume too high or too low a
plane* in the orbit.   To these conditions he has given the names
" *anaphoria*," " *kataphoria*," " *anatropia*," and " *katatropia*"
(see page 35).

For the determination of these conditions Dr. Stevens has
devised an instrument, known as the *tropometer* (Fig. 20),
by which the absolute arc of rotation inward, outward, downward, and upward can be scientifically measured, of either
eye independently.   As it is likely to prove of great value in
future work, it may be well to give a description of it and its
working.

This instrument consists of a very ingenious head-rest, by
means of which the head of the patient can be placed in the
same relation to the instrument at every sitting.   The eyes are
maintained upon the proper plane by this head-rest, because
the head can be adjusted accurately and made absolutely
immovable.

A wooden bar for each patient is first slipped into the
head-rest of this instrument.   Upon this the teeth of the upper
jaw of the patient rest, while the forehead is pressed tightly
against a supporting band of iron and retained there by an iron
loop screwed tightly to the occiput.   There are also two other
fixed points of metal in the head-rest, one of which should
touch the point between the eyebrows (glabella) and the other
the upper lip at its junction with the nose.

The importance of immobility of the head and the fixed
relationship of each eye to the barrel of the instrument during
the examination of the patient is vital to accurate measurement

of the various rotations of either eye that are to be measured and recorded by this instrument.

The instrument itself consists of a metal barrel fitted on one end with an adjustable eye-piece (into which the physician looks) and on the other end with a square box containing a forty-five degree prism. Opposite the center of this prism is a hole into which the patient gazes with the eye to be measured.

The opening for the eye-piece is at right-angles to the opening into which the patient looks; hence the barrel of the

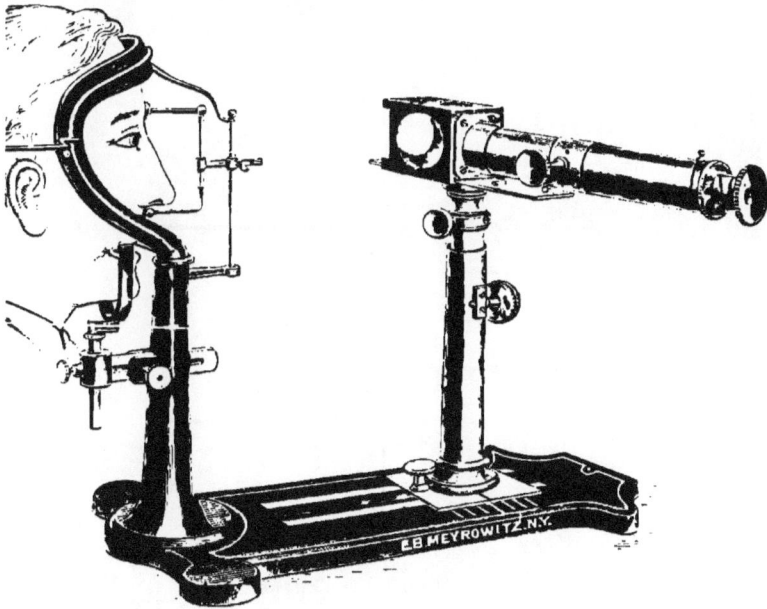

FIG. 20.—STEVENS'S TROPOMETER.

instrument stands at right-angles to the head-rest when in use, and the physician sits at the right-hand side of the patient. Within the eye-piece of this instrument there is placed a *carefully graduated scale*. To the center of this scale a reflection of the cornea of the eye under observation is first made to accurately correspond. The patient should be placed in a window, where a strong light falls upon his face, before the head is perfectly adjusted and made immovable. This bright illumination enables the physician to see the eye very clearly in the barrel of the instrument, after the eye-piece is properly

focused. The edges of the cornea are first made to correspond
to the two dark lines in the eye-piece by sliding the instrument
forward or backward on the stand and raising or lowering it by
the set-screw. The patient is then instructed to *look as far
upward* as possible, and the point on the scale is noted that
corresponds with the upper edge of the cornea (the movement
being reversed in the instrument). In the same way the patient
is instructed to *look as far downward* as possible, and the arc of
downward rotation is measured. In making the latter test the
upper eyelid must be raised by the examiner in order to see the

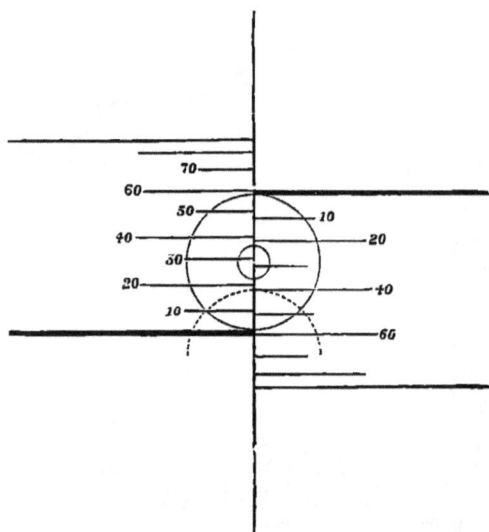

FIG. 21.—TROPOMETER SCALE.

edge of the cornea. The scale is then turned at right angles
and the arc of rotation, nasal and temporal, measured in a similar
manner. A number of tests should be made upon each patient
before any definite conclusions can be drawn from them, for,
until the patient fully understands what is required, a slight
effort to turn the head or a neglect to exert the full power of a
muscle may make the records inaccurate.

The instrument is yet too new to allow any one to state
positively the normal arcs of rotation; but the tests of a large
number of cases in my office seem to show that the following
figures are to be expected in the normal eye and that any marked

deviation from them is abnormal: *Upward rotation*. 32°; *down-ward*, 45° to 50°; *nasal*, 55°; *temporal*, 50°.

MEASUREMENT AND RECORDING OF THE VISUAL FIELD.

The perimeter is an instrument used to determine the ex-tent of the visual field.   There are various forms of instruments

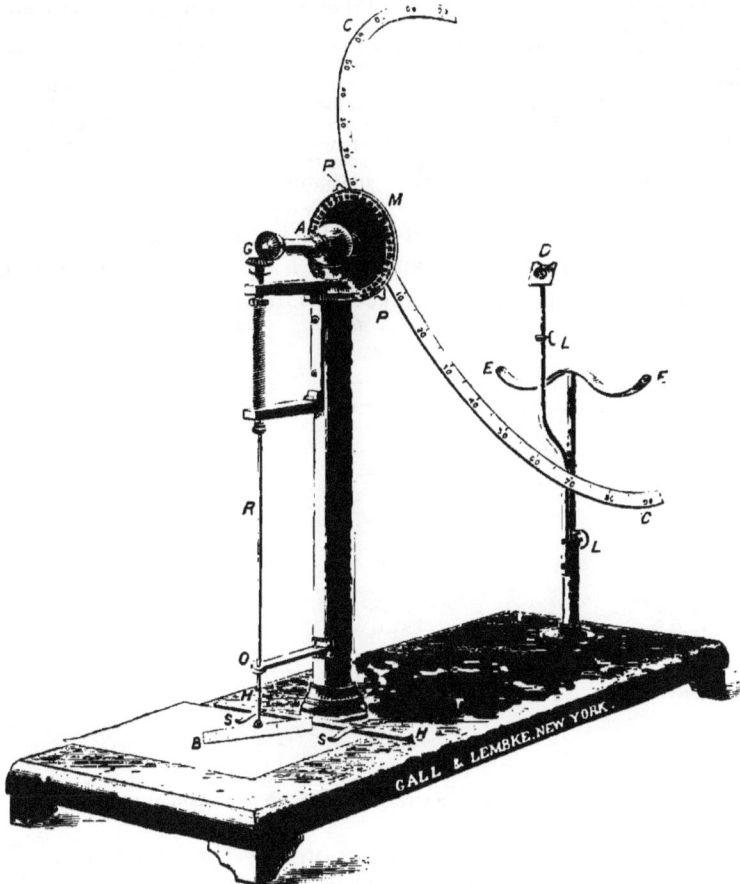

FIG. 22.—GILES'S PERIMETER.

The perimeter, illustrated by the accompanying engraving, differs from others in use chiefly in the device for recording observations, which is seen at the left of the engraving.   This device consists of a graduated index-bar (*B*), fixed at the lower end of a vertical rod (*R*), which is connected with the axis (*A*) of the in-strument by means of a beveled gearing (*G*) in such a way that every change in the position of the graduated arc (*C*) is indicated by a corresponding change in the position of the index-bar.

The instrument is designed to be used with Emerson's perimeter charts, but any other chart may be used. or. if no chart is at hand. a blank sheet of paper will serve the purpose. because the angle of the graduated arc is always shown by the position of the index-bar; when the Emerson chart is used. it is to be placed under the index-bar with the base toward the patient and against the brass guide (*H*). There it is held by means of the spring clips (*S, S*). In this position every portion of the field as represented upon the chart will be opposite the corresponding portion of the field of the eye under observation, and consequently there will be no danger of recording an observation upon the wrong side of the chart and thereby reversing the field.

The axis (*A*) is hollow ; so that, if desired, the observer may keep watch of the patient's eye through the opening. At the axis is a graduated circle (*M*), the position of which is fixed, while around it revolve two pointers (*P, P*), which are fixed to the arms of the arc.

The readings shown on the graduated circle should correspond exactly with those shown upon the chart by the index-bar. By means of this additional index the observations, as taken upon the chart, may be verified ; and, if by accident the index-bar is moved from its proper position, the true position may always be found by a comparison of the two scales. If the recording apparatus should be broken by any accident, the degrees could be read upon the graduated circle and noted upon the chart in precisely the same way as with perimeters of the ordinary form which are not provided with a recording apparatus.

devised for this purpose, but the best consists of a graduated arc, revolving about the center of a small graduated circle. The chin of the patient is placed upon a rest, adjusted so that

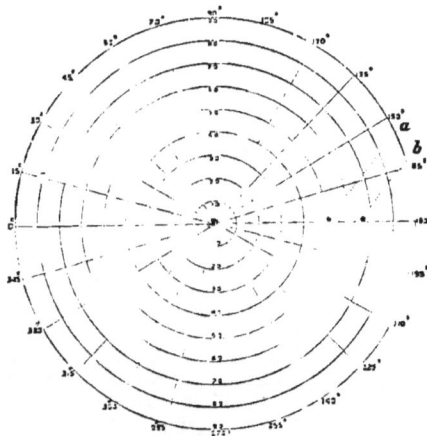

Fig. 23.—PERIMETER CHART, ADAPTED TO ALL PERIMETERS.

the eye to be examined is at the center of the large arc. The vision is then fixed upon the center of the pivot upon which the arc moves. This is often hollow, to enable the examiner to see that the eye of the patient does not wander. A small, white disc[1] on the end of a rod is then moved along the arc from without inward until observed by the patient, and the point at

[1] A red, blue, or green disc may be used to measure and record the visual areas for these colors when deemed important.

which it is first seen is recorded on a printed chart, arranged to correspond to the scale on the perimeter. The arc is then moved to each meridian in succession, and the extreme point of vision marked in each, until the circle is completed.

. A line joining these points on the chart will define the " field of vision" in the examined eye, the other eye being covered during the test.

The extent of the visual field in health will depend somewhat upon the formation of the face, as a prominent nose, overhanging brow, or high cheek-bone will limit the field to some extent. The white areas in Fig. 24 give the average limits of the normal visual area of each eye.

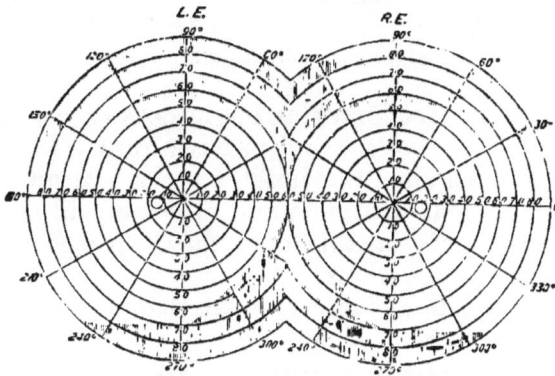

FIG. 24.—EMERSON'S PERIMETER CHART, ADAPTED TO ALL PERIMETERS.

The ordinary physiological limits of the field are about as follow: Outward, 90°; outward and upward, 70°; upward, 50°; upward and inward, 55°; inward, 60°; inward and downward, 55°; downward, 72°; downward and outward, 85°.

The later types of perimeter have an *automatic-registration attachment* that saves the operator much time and trouble.

DIAGNOSTIC ATTITUDES OF THE HEAD AND PECULIAR FACIAL EXPRESSIONS CAUSED BY MALADJUSTMENT OF THE EYE-MUSCLES.

An expert in the testing for and correcting of anomalies in the muscles of the orbit is often materially assisted in making a diagnosis and deciding upon his line of treatment by the expression of the patient's face and certain peculiar and persistent attitudes of the head. In certain cases very complex eye-problems are made far more simple in their solution by a study of the face and posture of the head than the records of the case would at first indicate.

For example, some extreme cases of double convergent squint are not in reality due to any abnormal condition of the

FIG. 25.—EXPRESSION OF A PATIENT WITH A MARKED HYPERESOPHORIA.
The head is tilted, indicative of hyperphoria.

internal or external recti muscles, but rather to a *defective power of rotation of both eyes* upward or downward. In such cases any operative steps upon the inside muscles are apt to be negative in their immediate results and ultimately an injury to the patient. In patients of this type the tropometer materially assists in making a diagnosis.

There is no field in surgery to-day that is complicated with

FIG. 26.—EXPRESSION OF A PATIENT WITH LATENT HYPERESOPHORIA.
The tilting of the head (indicative of hyperphoria) is marked.

greater difficulties, and in which greater precautions against mistakes have to be taken, than in scientific attempts at correction of extremely "cross-eyed" patients.

The experiences of oculists, as well as those of the general practitioner, confirm this statement, as the percentage of accurate adjustment after squint-operations is not large and extreme overcorrections very common, even in the hands of skillful operators.

The few hints that are given here will, I trust, aid the reader.

(*A*) A PERSISTENT TILTING OF THE HEAD TO ONE SIDE usually justifies the suspicion that one eye tends to rise higher than its fellow,—the condition known as *hyperphoria*. The head is usually inclined toward the side opposed to the high eye, the explanation of this being that the high eye tends to see an image lower than it should be and the low eye an image higher

FIG. 27.—EXTREME MANIFEST HYPERPHORIA WITH SOME ESOPHORIA (LATENT).

The tilting of the head (that often accompanies hyperphoria) is quite marked.

than it should be. By tilting the head toward the side corresponding to the low eye, the patient is enabled to bring the images more nearly to the same plane and to fuse them with less effort.

The condition of hyperphoria in these patients may not be at once disclosed by the phorometer (Fig. 12) even when the head of the patient is placed in the correct position; but it should be diligently sought for on several consecutive days, and any marked difference in the power of sursumduction in the two

FIG. 28.—EXPRESSION OF A PATIENT WITH AN ENORMOUS AMOUNT OF LATENT ESOPHORIA.

eyes should always be carefully noted (see page 46). Sometimes, by the judicious wearing of prisms for a time with the base up or down, a hyperphoria that was at first totally latent may be discovered; in which case the difference between the sursumduction of the right and left eye will be double in degrees the amount of hyperphoria corrected by prisms.

I would caution the reader against too hastily operating upon the lateral muscles of a patient who has persistently carried the head to either side. The correction of the latent hyperphoria alone will often be followed by a marked improvement in all of the eye-tests, irrespective of what anomalies of adjustment the patient at first disclosed.

(*B*) In *typical cases of esophoria* there is a PECULIAR INTENSE EXPRESSION OF THE FACE. This is hard to describe in words. The open and frank countenance seems to be wanting; and in

FIG. 29.—EXPRESSION OF A PATIENT WITH EXTREME ESOPHORIA CLOSELY BORDERING UPON CONVERGENT STRABISMUS, BUT MAINTAINING BINOCULAR VISION.

its place comes an expression that tends to create rather the feeling of distrust.

When the esophoria is extreme and bordering upon a slight tendency to squint, it is extremely difficult for these patients to look steadily in the eyes of any one conversing with them for any length of time without looking off or dropping the eyes; hence the character of these persons is often misconstrued because of their physical disability.

FIG. 30.—EXPRESSION OF A PATIENT WITH A HIGH DEGREE OF MANIFEST ESOPHORIA AND A MODERATE DEGREE OF HYPERPHORIA (LATENT).

The head is slightly tilted,—a posture that is suggestive of existing hyperphoria.

This same tendency to look away is also observed in patients with a slight tendency to divergent squint.

(*C*) In *extreme cases of exophoria* the face of the patient has rather a PECULIAR BLANK EXPRESSION.

Very often one eye will be observed to wander off toward the outer canthus momentarily, when the patient is not looking very intently into the face of the examiner. The typical "wall-eyed " expression, however, is only observed in cases of extreme divergent squint.

(*D*) In cases of anaphoria (page 35) the FOREHEAD IS THROWN FORWARD AND THE CHIN DRAWN IN CLOSE TO THE CHEST.

This gives the patient the appearance of always looking out from under his eyebrows.

The necessity of pulling the eyes down to the normal position is greatly overcome by this attitude. It thus relieves the strain on the inferior recti muscles. This attitude of the head has great clinical significance.

(*E*) In cases of kataphoria (page 35) there is usually a marked tendency toward a PROTRUSION OF THE CHIN. This attitude allows the eyes to assume a plane much lower than the

FIG. 31.—UNILATERAL EXOPHORIA CLOSELY BORDERING UPON UNCONQUERABLE DIPLOPIA (CROSSED DIPLOPIA).

normal, and the superior recti muscles are greatly relieved by it.

THE USE OF THE OPHTHALMOSCOPE.

A little has already been said, in preceding pages, upon the use of this instrument for the purpose of excluding or detecting diseased conditions of the retina.

It remains for me, however, to state my views as to what I regard as proper limitations upon the use of this valuable instrument.

1. This instrument affords *only an approximate* and *often an erroneous estimate of refraction* (see page 50), even in the hands of those who have the greatest skill and experience in its use.

For many years in the past (and still by some oculists of repute) refractive errors were estimated and glasses prescribed by the use of the ophthalmoscope alone, simply because it saved the patient the annoyance of having the pupil dilated.

While such tests often impress the patient with a greater

ability on the part of the oculist than he possibly deserves. I am satisfied that those who have the greatest skill as ophthalmoscopists not infrequently make very serious mistakes in esti-

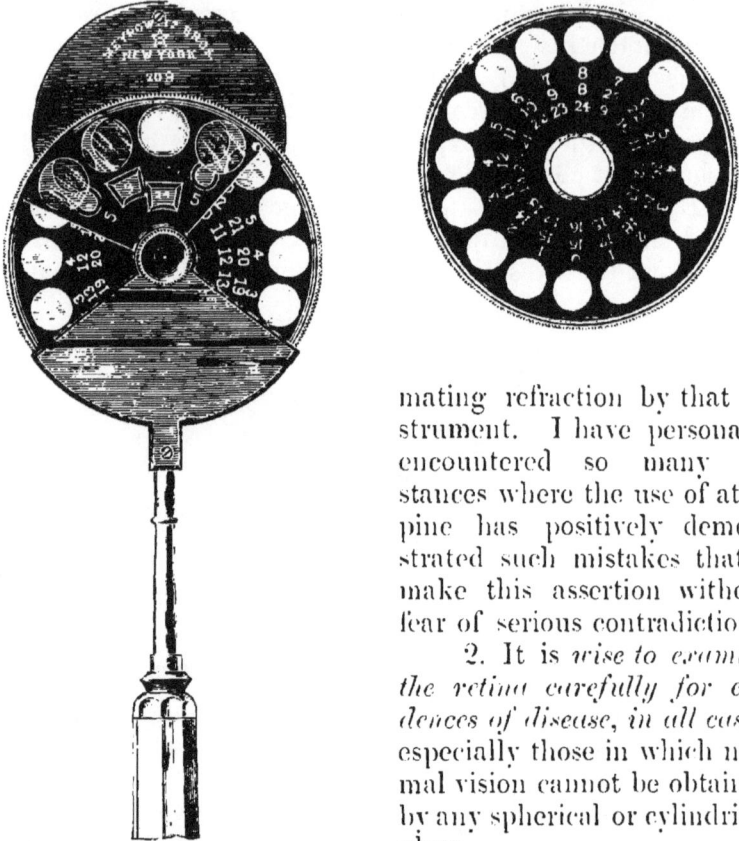

FIG. 32.—LORING'S COMPLETE OPHTHALMOSCOPE, WITH QUADRANT AND THREE ROWS OF FIGURES AND TWO CONDENSING-LENSES.

This instrument has a third row of figures, showing every combination up to + 23.0 D. and — 24.0 D., and obviates the necessity of mental calculation beyond 16.0 D.

mating refraction by that instrument. I have personally encountered so many instances where the use of atropine has positively demonstrated such mistakes that I make this assertion without fear of serious contradiction.

2. It is *wise to examine the retina carefully for evidences of disease, in all cases,* especially those in which normal vision cannot be obtained by any spherical or cylindrical glass.

The prognosis in any given case is seriously modified by the detection of organic retinal disease; and the presence of complicating kidney disease, that may not have been suspected either by the patient or by the physician, is sometimes brought to light by the appearance of the retina.

3. The ophthalmoscope is a valuable instrument in determining the *refraction of feeble-minded patients or of young*

children. Among these two classes the use of atropine prior to
the examination with the ophthalmoscope eliminates one element of possible error,—*i.e.*, the presence of ciliary spasm.

In *strabismic children* the determination of the refraction
of both eyes and the correction of all errors found by means of
properly fitted and centered glasses should always be done prior
to any surgical steps for its relief.

4. The detection of organic changes within the eye
(whether it be in the cornea, lens, vitreous, choroid, or retina)
necessarily modifies the prognosis (as before stated) even where
marked heterophoria exists; and, furthermore, in many instances
constitutes a bar to operative procedures upon the eye-muscles.

5. To be skillful with the ophthalmoscope requires, perhaps, *greater natural aptitude and experience* than with most
instruments. It is not every one who can control his accommodation at will while using this instrument,—a factor in such
examinations that may lead to serious error in the estimation
of refraction.

Most oculists of repute think that they can do this with
some approach to certainty; but I am satisfied that many fail
oftener than they think, and are led into unconscious error in
using this instrument. I have had many oculists tell me that
they had found ophthalmoscopic work more satisfactory to
themselves after they had lost their power of accommodation
from age than during their early professional life.

6. The choice of an instrument depends largely upon the
preference of the oculist. While the mechanical construction
may vary, the general arrangement and principles of all modern
instruments are practically alike.

Fig. 32 shows one of the most popular forms of ophthalmoscope.

It consists of a concave tilting mirror, perforated in the
center, so that the observer's eye can look directly at the eye to
be examined practically from the point of illumination.

Behind the mirror is a metal disc, which carries a revolving
set of concave and convex lenses. These can be thrown at
will behind the perforation in the mirror for the correction of
any error of refraction.

A biconvex lens of two and one-half inches' focus usually
accompanies each ophthalmoscope, for use in the indirect
method of examination.

There are two methods of using the ophthalmoscope,—called the "direct method" and the "indirect method."

In either mode of examination the position of the light is the same. It should be *behind the patient, on a level with the orbit*, and *slightly toward the side* corresponding to the eye to be examined. The light most frequently used is an Argand gas-burner; but electricity may be used if the globe is made of ground glass, so that the wires are not seen.

In the DIRECT METHOD (Fig. 33) the examiner sits in front

FIG. 33.—OPHTHALMOSCOPIC EXAMINATION. (De Schweinitz.)[1]
Method of the upright image. Observer and patient in the correct position.

and a little to the side of the eye to be examined, and holds the instrument in the right hand before his right eye in examining the right eye, and *vice versâ*.

With the eye of the physician placed close behind the aperture in the mirror, the light is first thrown upon the eye of the patient, who is directed to look over the shoulder of the examiner at as distant an object as possible. The pupil at once assumes a bright-red color, with any opacities of the cornea or lens appearing as dark spots upon the red background.

[1] Figs. 33, 34, and 35 from de Schweinitz on "Diseases of the Eye" (W. B. Saunders, Philadelphia). By permission of the author.

The eye of the examiner (*with accommodation relaxed*) is then brought closer to the eye of the patient, until the deeper parts of the eye, blood-vessels, optic disc, etc., are plainly seen.

In myopic and hypermetropic eyes the correcting lens must be placed behind the mirror before the details of the fundus can be clearly seen. Only a small portion can be seen at once, but, by directing the patient to look from side to side, upward and downward, every part can be brought into view.

In the INDIRECT METHOD (Fig. 35) of examination a bicon-

FIG. 34.—OPHTHALMOSCOPIC EXAMINATION. (De Schweinitz.)

Method of the upright image. Observer in an incorrect position.

vex lens is held in front of the patient's eye, when, with the head at a distance of about fifteen inches, an *inverted image of the fundus* is seen in front of the lens.

In this manner a smaller image is obtained than by the direct method, but a larger part of the fundus can be seen at once.

The image is magnified about five times in the indirect and fifteen to twenty times in the direct method, depending upon the refraction of the eye, the enlargement being greater in proportion to the depth of the eye.

Pages might be written upon the use of the ophthalmoscope, but the size of this small volume will not permit of it. Long-continued practice alone, under a competent instructor, can make a novice proficient in its use.

A novice should begin by practicing with this instrument in throwing the light with rapidity, certainty, and steadiness upon various fixed points, such as buttons on the clothing of a subject placed in front of him, and various portions of the face, with the light in proper position; then by keeping up a steady illumination of the retina while the eyeball of the patient is slowly moved in all directions; then by attempts to get a

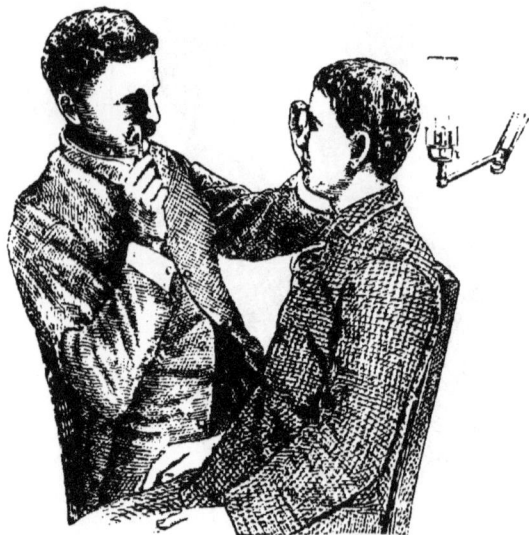

FIG. 35.—METHOD OF THE INDIRECT EXAMINATION WITH THE OPHTHALMO-
SCOPE. (De Schweinitz.)

clear image of the *retinal disc* of the patient with a normal pupil.

The cuts introduced here will doubtless give the reader a better general idea of the relative positions of the examiner, patient, and point of illumination during an ophthalmoscopic examination than the preceding description.

In closing this chapter the author feels that much has doubtless been omitted that might prove of material aid to a novice in testing the eyes for errors of refraction or anomalies of adjustment of the muscles of the orbit; but it is impossible in print to give all the suggestions that clinical work may at times

justify, and that experience alone can teach even to those who
have given years of patient labor to this field.

The expert sees at a glance often a suspicion of trouble
that it may take many weeks or months to positively demonstrate.

Again, by intuition, he frequently sees and may rapidly
find a solution of certain complex eye-conditions that have
escaped recognition in the past by oculists who have ignored
the existing heterophoria.

The time is near at hand to-day when scientific men will
refuse to be satisfied with a cursory examination of the eyes of
any patient for astigmatism, myopia, or hypermetropia and a
glance at the retina with the ophthalmoscope. This does not
now, as it has done in the past, warrant any oculist in dismiss-
ing his patient with an assurance that "the examination is complete and final." Such a method of procedure (common as it
has been in the past) is not complete, nor should it be allowed
to rest as the basis of a final verdict upon a point of vital importance to thousands of sufferers from nervous disturbances.

CHAPTER III.

EYE-STRAIN AS A CAUSE OF HEADACHE AND NEURALGIA.

With Remarks Concerning the Prevalent Fallacies Respecting these Conditions.

Some time since I encountered in my reading the following extract from the pen of a great French philosopher: "Nature fights with disease a battle to the death. A blind man armed with a club—that is, a physician—comes in to make peace between them. Failing in that, he lays about him with his club. "If he happens to hit disease he kills disease; if he hits nature he kills nature."

There is in this charming piece of satire more than a single atom of truth. In nothing is it better illustrated to-day than in the literature of medicine relating to neuralgias and headache. There are undoubtedly more sufferers among the community at large from these two forms of paroxysmal pain than from any other morbid conditions. This statement is, I think, sustained by all carefully prepared statistics. The amount of literature which has been contributed to this subject would also indicate to the candid reader not only their great prevalence, but the peculiar intractability, in some cases, of these types of pain in spite of drugs.

A very large percentage, moreover, of victims to headache and neuralgia fail to disclose any positive evidences of actual disease, in spite of the greatest care in searching for a cause. The affected nerve-trunks, the brain itself or its coverings, the heart, the lungs, the kidneys, the liver, the stomach, and the intestine, etc., seem in most cases to be capable of performing their normal functions, save when some temporary disturbance (not generally referable to a very definite cause) occurs simultaneously with, or prior to, one of these paroxysmal attacks of pain. Such disturbances are too frequently regarded by various observers as of great significance, but to my mind clinical experience teaches us that they are to be classed rather as mere coincidences, or as a part of the attack itself.

We, as a profession, have yet much to learn respecting the

remote effects of a disturbed nervous equilibrium upon the organs of the chest and abdomen. We have also, in the light of modern physiology, much to unlearn, many inherent prejudices to overcome, many clinical conclusions to discard (that have grown upon us from defective habits of reasoning) before our patients can derive the full benefits of our own advancement.

If we start out, for example (as many physicians do), with the premise (generally an incorrect one) that every patient who has a disturbed stomach, a sluggish bowel, or clay-colored fæces prior to, or in connection with, a headache owes his headache directly to his stomach, liver, or bowel, and must base his recovery upon some drug which shall re-establish their action, many patients will become imbued with ideas which may, perhaps, indefinitely postpone, or totally preclude, the opportunity of obtaining permanent relief by the removal of a cause not affected by drugs and too often unsuspected by the patient.

Too many medical men ignore the fact to-day that an *underlying cause exists* in almost every case of intractable neuralgia or headache.

Yet they speak with assurance respecting the clinical significance of some slight draught of air; look for some trivial excess or error of diet as an important factor; condemn hastily, and on general principles, some acquired habit which has been tolerated by the patient for years without apparent injury (such as tobacco, for example); shift the responsibility from themselves upon that much abused organ the liver; and in countless ways strike wildly about with their club of unscientific deduction (according to the French philosopher), hoping to kill a disease whose causes and nature they may not have very carefully investigated.

A bad tooth, impacted wax in the ear, eyes with defective refraction, imperfectly balanced eye-muscles, an obstruction to the passage of air through one or both of the nasal cavities, some hidden rectal irritation, and many forms of pelvic disease too often remain unrecognized in many patients as factors in causing headache and neuralgia, while learned dissertations and useless medication are offered to the patient as a means of relief.

Hardly a day passes that I do not see some sufferer of this class in my office, who has passed through many years of periodical suffering from frequent headaches or persistent neuralgic attacks unnecessarily, simply because the exciting cause has either never been sought for or investigated by imperfect methods and with too little care.

Headaches and neuralgias may be discussed under three prominent classes,—the *toxic, organic*, and *reflex varieties. Of these the latter is the most common.*

Toxic Headaches.—There is, unquestionably, a certain proportion of individuals (where headache is a marked symptom) that owe their suffering to some abnormal condition of the blood. Among this class of causes we must recognize the poisoned blood of Bright's disease, the saccharine blood of diabetes, the decrease of red and white corpuscles observed in anæmia and leucocythæmia, the blood-changes produced by malarial poisoning, and some typical cases of poisoning by lead, mercury, arsenic, alcohol, etc.

It must be acknowledged, however, that sufferers from these forms of headache are comparatively infrequent, when contrasted with the thousands in which no evidence of actual disease can be discovered by scientific means. Yet sufferers of this class are too commonly encountered to justify the omission of a careful examination of the urine in any individual case,—an excretion whose constituents decide quite conclusively the existence or non-existence of any typical morbid state. If deemed necessary, the blood also, and other fluids of the body, can be subjected to scientific investigation. It is certainly a culpable oversight, in any case where headache exists, to neglect a careful analysis of the urine.

Headaches and Neuralgia from Organic Disease of the Brain, Spinal Cord, or Nerves.—Again, there is encountered a class of patients who suffer from more or less obstinate and intractable headache and neuralgia as a result of organic disease of the brain or its coverings.

Certain forms of spinal disease are likewise apt to create more or less excruciating forms of pain that might easily be, and too often are, mistaken for neuralgia in some of its forms. These do not, however, as a rule, cause headache, if the spinal nerves are alone involved.

In the cerebral and spinal types of cases the ophthalmoscope becomes a valuable instrument, since it will often show alterations in the retina or " disc." Moreover, other symptoms referable to the power of motion, sensation, or the special senses are very apt to exist and to point directly to the presence of organic disease of the nerve-centers.

While it is advisable, therefore, to exclude organic disease of the nerve-centers by appropriate tests employed for that purpose in every case of intractable headache or neuralgia, still it may be said that such cases constitute but a small proportion of those who seek medical aid for the symptoms under consideration in this chapter.

Admitting, then, the fact that the presence of headache and neuralgia may, in some instances, be but a clinical evidence of organic disease, it is our manifest duty in every case to exclude, as far as it is possible to do so, the existence of organic disease in each individual case ; or, if such be detected, to combat it by wise therapeutic measures. On the other hand, the comparative percentage of such cases is too small, and the possibility of other causes (as co-existing factors) too large, to justify us to-day in neglecting to take into account in any case the various types of reflex disturbance to the nervous centers which clinical experience has shown may both modify and cause severe cranial pain or neuralgic paroxysms.

REFLEX HEADACHE AND NEURALGIAS.—I have deemed it wise to insert these few remarks as a preface to this chapter for the following reasons :—

1. Because the profession at large has not, in my opinion, woke up thoroughly to the fact that a very large percentage of headaches and neuralgias can be relieved, and often radically cured, without drugs.

2. Because most of the later contributions to this subject have very clearly demonstrated that the authors of these papers are singularly apathetic to facts that have already passed beyond the limits of unsupported theory, and are being recognized by many of the eminent scientists of our day.

3. Because remarkable cures of headache and neuralgia are being daily forced upon professional notice, which have been brought about solely by the removal of some source of reflex irritation,—such, for example, as decayed or ulcerated teeth, a deflected nasal septum, and errors of focus or of the muscular adjustment of the visual apparatus.

4. Because any new therapeutic measure meets with far less opposition than does the introduction of a scientific method of search for obscure causes of disease. The public has been

educated to want drugs and doctors to rely too much upon them.

5. Because many of the older views respecting the underlying factors of disease are destined to be overthrown by many of the later discoveries in physiology, bacteriology, etc.

6. Because all views not in accord with popular belief are constantly being misrepresented, misunderstood, and misapplied by those who either do not wish to know them (from personal prejudice) or who have had little practical experience in the methods advocated. The history of medical progress is replete with examples of violent opposition to all innovations,—such, for example. as vaccination, listerism, ovariotomy, etc.

7. Because scientific methods of precision require a special training of mind and eye and hand not commonly required of undergraduates in our medical schools.

Not long since I received a letter from a physician (who had asked of me some suggestions respecting the ocular treatment of a case that had proven intractable to all medical agents) in which he says: "What troubles me now is the fact that I am not sufficiently posted on the diagnosis of the eye to use your valuable information. I am very much inclined, however, *to go it blind on the case I spoke of, and give it a trial.*"[1]

Such an evident misconception of the experience, skill, and care required in the diagnosis and treatment of an error of refraction or a defective muscular adjustment of the eyes requires no further comment from me.

For some years past it has been my custom to examine repeatedly, and with care, the visual apparatus of every patient sent to me for the relief of headache and neuralgia. I have already published, from time to time, many cases where this line of research has been followed by marked and permanent benefit.[2]

I deem it worthy of remark, in this connection, that less importance is being attached to-day than formerly to the clinical determination of modifications in the calibre of the blood-vessels of the brain. The conditions known as "cerebral congestion" or "cerebral anæmia" of certain authors who have written extensively upon headache and neuralgia are very often the results

[1] Italics my own.
[2] New York Medical Journal, January, 1888; Lectures on Nervous Diseases. Philadelphia, 1888 (The F. A. Davis Company, Publishers); Medical Register, Philadelphia, November 19, 1887.

of underlying factors. Their existence (as has been claimed) may sometimes be revealed by the ophthalmoscopic examination of the vessels of the retina and confirmed by the effects of nitrite of amyl upon the patient; but when so, they are probably to be regarded rather as an evidence of a functional derangement of the vasomotor system of nerves than as permanent factors in headaches or neuralgias.

I have known many patients who have followed, with negative results, a prolonged course of treatment (by ergot, bromides, amyl, etc.) which was based upon the examination of the retinal vessels; and I have often seen them recover from their headaches and neuralgias without drugs when an anomaly of refraction or of the muscular adjustment of the eyes was corrected.

I would not be construed as denying that the blood-vessels of the brain might not have been abnormally dilated or peculiarly contracted in many of these patients at the time when their retinal vessels were examined; nor would I utterly reject the hypothesis that the retina sometimes affords us a valuable means of determining, by our sense of sight, the condition of the cerebral vessels in any given individual.

What I do mean to assert is this: That changes in the cerebral circulation (similar to those observed in the case of a blush upon the cheek) may be caused by subtle nervous influences that ergot, bromides, nitrite of amyl, or other drugs will not arrest; that a classification of headaches or neuralgias which is based upon so variable a sign is unscientific; and that any line of medication which is directed toward this condition alone is very liable to be unsatisfactory both to the physician and his patient, sooner or later.

To illustrate this point I will mention a remarkable experience of my own that impressed me strongly at the time. Some eighteen years ago, when almost in despair from continuous and intractable headache, and doubtful of my ability to long endure it, two oculists of equal prominence and ability were asked by me to examine my own retinæ by means of the ophthalmoscope. One diagnosed my condition as "typical nicotine poisoning of the retina"; the other discovered what seemed to him to be conclusive evidences of "congestion of the brain," that nothing but prolonged rest from work, in his opinion, could relieve. Neither suggested the detection of any latent refractive error or the use of glasses. My sight was apparently perfect and unusually acute. Subsequently the instillation of atropine into my eyes (which was used only at my earnest solicitation) and the correction by proper convex glasses of a hypermetropia of 3.50 diopters (not previously suspected) restored me to health and comfort as if by magic. A complete and immediate cessation of all pain for over nine months followed the correction of my refractive error; and for many years I have been almost entirely free from pain, in spite of continuous eye-work at my desk and elsewhere.

I can point to case after case in my professional experience, since my own recovery, where the cause of neuralgic attacks, excruciating headaches, vomiting, extreme nervousness, and many other symptoms (not apparently connected with eye-defect) would have remained unrecognized if atropine had not been employed. As I have said on a previous page, " patients who boast of their acuteness of vision, and who apparently justify their statement by reading test-type at a distance, without the aid of glasses, are often astonished and sometimes alarmed, at the immediate loss of this power which is brought about by the use of atropine. This surprise is heightened when (by the use of proper lenses) their power of vision for distance is immediately restored, and they become conscious, for the first time, of the muscular effort which they have been compelled, in the past, to exert in order to see without them. I shall never forget, personally, the sensation which I experienced of 'seeing without effort' when latent hypermetropia was discovered in my own eye and corrected by glasses.

" These experiences are well-known facts among oculists, but to the profession at large they often occasion as much surprise as to the patient."

A report of fifty consecutive cases of typical headache and neuralgia, taken from my private case-book as confirmatory of the views that are here advocated, are quoted, with some later reports added, from a previous *brochure* published in 1889.

Space will not allow of the full records of all of these cases. I have tabulated the main points of each case[1] and given histories of those only that presented some particular difficulties either in diagnosis or treatment. These will be

---

[1] In the table appended to this chapter I have only noted the refraction as observed when the eyes were under atropine.

numbered so as to correspond to the table. In such a tabulated record it is impossible to show many fluctuations observed in the muscular tests during treatment.[1]

Many additional cases where headaches were a prominent symptom are recorded in subsequent chapters (see sections IV, VI, and VII), but other symptoms of a graver kind constituted the chief clinical features.

CASE III.—This case is one of special interest. Both the patient and her sister had been delicate for many years. The patient had never menstruated; had been addicted to " fainting-spells," which occurred frequently and at the most inopportune times, so that she would fall unconscious in the street, etc.; had suffered from melancholia, so that once she attempted suicide; had been a victim to nervous dyspepsia; and had suffered more or less from asthenopia.

Her family history was a bad one, showing a marked hered-ity to nervous derangements. The mother had the same " faint-ing-spells "; the maternal uncles all died of diabetes; one cousin of the patient has epilepsy; and two other cases of epilepsy existed in her paternal ancestry.

The eye-tests showed unequal refraction in the eyes; the right eye showed a decided tendency to assume a higher plane than the left, although she could fuse her visual images, even with a red glass before either eye.

As she was unable to desist, for a time, from an occupation demanding the use of the eyes at the near point, the manifest hyperphoria was corrected (within 1°) by a prism. The effect of the prism has thus far been very beneficial.

That this defect in the vertical movements of the eyes is a congenital one I have had an opportunity of quite positively demonstrating by similar examinations made upon others of her immediate family.

The so-called " fainting-spells" to which this patient was subject are probably closely allied to epilepsy,—a disease to

---

[1] Every patient examined for defective equilibrium in the eye-muscles instinctively strives (not by mere volition) to get binocular vision under the nearest approach to physio-logical conditions of which he is capable.

We are forced to admit, therefore, that what we detect in any patient and record as an error is, in reality, only what the patient cannot conceal, not necessarily all the defect in the muscular adjustment that actually exists.

Because a patient can momentarily perform a feat of eye-balance which approaches the normal state by the aid of his reserve-power, it is by no means proved that the eyes are habitually in equilibrium.

which she is liable by heredity. It is not common for faintness to prostrate people suddenly while crossing a street (without apparent cause or warning), and to be associated with a total loss of consciousness for quite a prolonged interval.

CASE IV.—This patient presented a peculiar heredity to nervous derangements. Insanity, epilepsy, and headaches have been developed in the ancestral lines. The patient has had attacks of marked petit mal, in which consciousness would be suddenly lost for intervals of from five to fifteen minutes. His eyes began to cause him distress about two years before I saw him. His mental faculties had at times been rendered peculiarly sluggish; so that the patient felt unable to handle complex business problems.

CASE V.—This case illustrates a peculiar instance of marked relief from a prostatic neuralgia (which bids fair to entirely disappear without local treatment) after the cessation of eye-strain. Headache and sleeplessness also existed as sources of great distress.

I have previously reported a remarkable case of this type of neuralgia (*New York Medical Journal*, January 7, 1888), which bears upon the correction of an eye-defect to such an extent that I quote it in full, as follows:—

"Male, aged 23, unmarried.

"*Family History:* The mother of the patient suffers from neuralgia and headache. The paternal grandfather had paralysis. The paternal grandmother was 'extremely delicate.' One brother suffers from headaches. Another brother is very excitable and of a highly-nervous temperament. No case of consumption has ever occurred in any branch of the family.

"*Eye-defects:* Hypermetropia (latent) of 2.50 D. Esophoria (manifest), 4°; subsequently 12° were elicited prior to any operative procedure.

"This patient had been under medical care for many months for a prostatic neuralgia, and had derived no benefit from local or general treatment. He developed melancholia, and would frequently retrace his steps for several blocks, during a stroll, in order to touch some object which he felt he should have touched when he passed it. The use of his eyes intensified his mental symptoms markedly. He also suffered from morbid fears. He had never had venereal disease. After partial tenotomies were performed upon his interni and his hypermetropia was corrected by + 1.50 spherical glasses, his recovery was very rapid and complete. He has had no abnormal mental symptoms or neuralgia of his prostate since the first operation (now nearly ten years). His father, one brother, and a sister have since been examined by me, and all had very marked eye-defect."

I have, at the present time, a third case under my care, which I think belongs to the same class of reflex neuralgias.

In every one of these three cases the relationship between eye-strain and the prostatic neuralgia had never been suggested to the patients by any of their numerous medical advisers or by their eye-symptoms. All had tried local treatment, directed to the prostate gland, without satisfactory results, under skillful and experienced hands.

CASES VI AND VII.—These were typical cases of severe paroxysmal pain, in which very gratifying results were quickly obtained by a graduated tenotomy of one muscle. Both had tried drugs without benefit.

CASE VIII.—This patient (the wife of a prominent divine) was sent to me from a distant city as an "incurable case of sick headache." The attacks had recurred so frequently and with such severity as to practically unfit the patient, for many years, for any social or household duties. All medicinal treatment had proved of no avail. No trouble had ever been experienced with the eyes, as far as the patient could recall.

Under atropine an astigmatism ($+ 1.00^c$) was discovered and corrected by cylindrical glasses. She also manifested within a week nearly 6° of right hyperphoria, and an esophoria which fluctuated from 2° to 8°. Prisms of 5° (in total) for her hyperphoria were worn by her over her cylinders for more than three weeks, during which time her customary headaches failed to appear and her apparent esophoria was favorably modified.

The vertical deflection was then corrected by a graduated tenotomy of the left inferior rectus.

A report received from the son of this patient states that " mother is almost entirely free from pain and has had no attack of sick headache for many months." The physical condition of this patient has improved to such an extent as to render her capable of daily feats of exertion that had heretofore been deemed impossible.

CASE IX.—This lady was sent to me by an oculist of prominence in the West, because he could not interpret to his satisfaction certain points of apparent conflict in the tests relating to muscular equilibrium in the orbits, nor could he cure her severe asthenopia and intractable neuralgia by any form of glass.

On the first examination any attempts to measure the sursumduction, adduction, or abduction of the eyes caused the most extreme faintness and nausea ; so much so that they had to be desisted from at that time. The patient disclosed, however, after some days of careful observation, a high degree of hyperphoria, with esophoria of 14° (causing homonymous diplopia with a red glass).

The correction of her astigmatism by glasses and the relief of the abnormal muscular tension in the orbits by graduated tenotomies on three muscles have greatly improved her condition ; but I am inclined to doubt that her deviating tendencies in the orbit have yet been completely rectified.

Her neuralgia and headaches had been lessened in frequency and severity at least one-half, and her asthenopia was decidedly diminished when she last wrote me. Unfortunately she is forced to use her eyes constantly at the near point.

CASE X.—This patient was sent to me from a distance by her physician for relief from terrible sick headaches that occurred once or twice each week. She also was unable to use her eyes for reading, sewing, or any prolonged effort without a severe headache occurring within twenty-four hours. She had been confined in bed for months, at one time, from an attack of nervous prostration.

She disclosed a high degree of hyperphoria and esophoria. These abnormal conditions seem at the present time to have been satisfactorily rectified by three graduated tenotomies. In one of her last communications to me she says: "To say that I feel better would be expressing it in very mild language. I feel perfectly lovely and can stand almost anything."

CASE XI.—In the ocular conditions of this patient a most complex anomaly of adjustment of the eye-muscles was encountered. The heredity of the patient was a particularly bad one.

Phthisis and epilepsy had developed in both the maternal and paternal ancestral lines, and one brother had developed epilepsy (with attacks closely bordering upon insanity) before the age of 10 years.

Two graduated tenotomies have thus far been performed upon the eye-muscles with happy results both as to the expression of the eyes and the frequency of headache; but the results are not yet what I believe they would be were the conditions encountered perfectly rectified. Such conditions as were observed in this case require time, patient observation, and extreme care in treatment.

CASES XII, XIII, AND XIV.—These were simple cases to diagnose and combat. All were treated by me by graduated tenotomies of the eye-muscles for some manifest anomaly. The improvement noted in all has been gratifying to the patients as well as to myself.

CASE XV.—This young lady was sent to me by a prominent New York surgeon, under whose care she had been, and at whose advice a plaster jacket had been worn continuously for a year with the hope of controlling a constant pain in the spine between the shoulders and in the lumbar region. She had also been under the care of one of the leading New York gynæcologists, who had helped her somewhat she thought, but had not relieved her head or spine. Her physical condition was one of extreme and progressive debility. Over two weeks elapsed before she could endure a carriage-ride to my office, after a letter of introduction had been given her, although she was living but a short distance from it, and the weather was favorable.

She had, when I first saw her, constant headache and spinal pain; was unable to read, write, or sew; and could sleep only a small part of each night. The condition of her eyes proved to be a very complex and difficult problem to solve and rectify. She disclosed, after patient observation, a high degree of hyperphoria (almost totally latent) and also a homonymous diplopia (when a red glass was placed before either eye). Repeated graduated tenotomies were performed upon the eye-muscles within a period of three months, when she returned to her home in a Southern State. Before she left New York she was walking about three miles each day; had experienced a great relief from her headache and spinal pain, and was better than she had been for many years. In one of her written communications, lately received, she says: "I am sleeping better and the pain between my shoulders is decidedly better. I have less pain in the back of my head. I drive, walk, entertain company, and go to church every pleasant day; so I know I must have gained in strength even since you last saw me." This case very closely resembles one reported by me in 1888, which will bear repetition here. My record of this case is as follows:—

"*Cerebral Neurasthenia, Constant Pain in the Head of Five Years' Duration, Asthenopia, etc.*—Female, unmarried, aged 21 years.

"*Family History:* Maternal aunt and five paternal relatives died of phthisis; two cousins had chronic chorea.

"*Eye-defects:* Patient had hypermetropia (latent) of 1.25 D. and exophoria (manifest) of 2°. A latent hyperphoria of 2° was subsequently discovered.

"This young lady was brought into my office by two assistants, who were obliged to carry her from the carriage. For several years she had been carried daily from her room to the library of her father's house, and, after reclining in a chair for a few hours, she would again be carried to her bed-room. She could manage with difficulty to walk slowly across a room. She had not been able to write, read, sew, or see her most intimate friends for five years on account of a constant pain in her head, which was rendered intolerable by any use of the eyes or excitement. Her symptoms began while at boarding-school, from which she was removed to her home in a recumbent posture and by easy stages.

"I used static electricity upon this patient for some weeks with a slight improvement in her power of walking, but no relief to her head. I then persuaded her to consent to a relief (by partial tenotomies) of her abnormal eye-tension. Tenotomies were then performed upon her left superior rectus and both externi within the space of two weeks. From that date her improvement was very rapid. She was sent home a few weeks later practically cured."

A letter from her physician, lately received by me, says: "Your patient is the wonder of this region. She rivals the 'Jersey Lily' in her feats of walking."

Before this patient was sent home she ascended and descended five flights of stairs daily, and averaged over a mile's walk each day without a companion to assist her. I would add to this record that this patient lately called at my office and informed me that she was studying physical culture, with a view of teaching it.

CASE XVI.—This case is one of the most puzzling problems that I have yet encountered. The patient is a young lady who had suffered from a constant pain over the eyes that nothing would relieve. She had been under the best medical care, and had consulted a very prominent oculist, who advised "a weak reading-glass." Her pain still remained unabated. For a long time after I had the opportunity of testing her eyes she failed to disclose a left hyperphoria, which her expression justified me in suspecting. Prisms were worn by the patient for this defect as fast as she revealed it (always keeping about 1° behind what the patient actually disclosed), until a total of nearly 8° of hyperphoria was thereby revealed by the patient. Two graduated tenotomies were performed to correct this deflection (the left superior and the right inferior muscles being button-holed). Both muscles were found to be enormously hypertrophied. The tendons were as thick as heavy buckskin. Subsequently both of the externi were operated on for an exophoria that had always manifested itself. The results thus far have been that the patient is greatly relieved from her headache, but is not absolutely cured. Her muscular balances seem to be nearly perfect.

CASE XXI.—This case has not progressed as well as I at first anticipated it would. The patient's eye-tests at the first interview led me to hope for marked and immediate benefit, but the results thus far are not as rapid as I expected. The patient was obliged to return to her home in the western part of this State, and I placed her under the care of an oculist who is quite well versed in the treatment of maladjustments of the eye-muscles. Before she left New York I had reason to suspect the existence of a latent hyperphoria.

CASE XXIII.—This girl weighed, when I first saw her, but ninety-two pounds, although five feet five inches in height. She was in a nervous condition that justified alarm, as morbid fears and emotional disturbances were becoming very pronounced. Her headaches were peculiarly distressing and very frequent. She had been under the care of several prominent physicians and one oculist of note. A weak cylinder had been prescribed for the right eye, to be used while reading' or sewing. A latent hyperphoria of 4° and an exophoria of 5° were corrected by three graduated tenotomies. She is now better in health than for many years past, and has gained some ten pounds in weight.

CASE XXIV.—This patient is a physician. Since childhood he has had a peculiarly "weak stomach" and neuralgic headaches of a severe type. Without a known exception his maternal ancestry for several generations have been similarly affected. His mother and one brother died of brain disease. Two sisters are victims to excruciating headaches. His father and paternal ancestry suffered more or less from sick headaches. Within the past fifteen years the patient has had frequent attacks of vertigo and more or less "confusion of thought" at times. Insomnia has also become almost a habit. The beneficial results of eye-treatment have been very marked in this case. He had esophoria of a high degree, and some left hyperphoria, to relieve which I performed three graduated tenotomies. The last report from this patient would indicate that he is in better health than at any previous period. He expresses himself as particularly grateful for the improvement in his headaches, digestion, and ability to sleep. He can now read, travel in the cars, and do gynæcological work without distress in his head; all of which used to be particularly difficult for him to attend to without entailed suffering.

CASE XXV.—This is particularly of interest from the fact that the patient had been afflicted for six years with very frequent paroxysms of neuralgia over the eyes, of an excruciating type, accompanied by vomiting, that were apparently due to diabetes. Dr. George T. Elliott, of New York, examined the urine several times while the case was under treatment, and noted a steady decrease in the percentage of sugar, without the aid of drugs or restricted diet,—from 3 per cent. to one-third of 1 per cent. The patient showed a high degree of left hyperphoria and esophoria. Graduated tenotomies were performed for the relief of the hyperphoria only, and for quite a long time all the neuralgic paroxysms ceased, and the improvement in the patient's condition was very marked. Unfortunately, the patient has thus far failed to allow me to complete the restoration of the other deviating tendency. I have received a letter from him in which he states that he has discontinued glasses and that his headaches have returned. I have advised him to return to New York and complete the line of treatment commenced. His urine has been repeatedly tested by his physician, and he has found no trace of sugar at any time.

CASE XXVII.—All efforts made thus far to relieve this patient of headaches have proven of no avail. No evidence of any organic disease has been detected by any of the physicians who have had him under their care. The heredity of the patient was bad, and justified me in hoping for better results. Phthisis was common in the paternal line. The mother had hemiplegia. A sister (now under my care) has very severe attacks of catalepsy and hystero-epilepsy. Twice in her life she has been supposed to be dead, laid out for burial, and the clothing prepared. This case was under the care of Dr. C. L. Dana, of New York, for quite a long time. He obtained negative results from electrical treatment and drugs. At my suggestion (when the apparent eye-defects had been satisfactorily overcome without any decided benefits) this patient placed himself under the care of Dr. J. W. Wright, of this city, for the relief of an anal fissure and hæmorrhoids. No marked change in the headaches followed the removal of this source of reflex irritation. As he had also a very marked deflection of the nasal septum, I advised him then to have that corrected. This step was undertaken, with admirable success, by Dr. D. Bryson Delavan, of New York, who removed quite a large piece of the septum. For awhile these steps proved of no material benefit as regards the headache, but he has lately reported that he considers himself as cured. I encountered a case exactly similar to this, some two years ago, which was sent to me from San Francisco. In this case little material benefit was obtained. It is still possible that some latent cause for eye-strain would have been detected by further observation.

CASE XXIX.—This patient had been under constant medical supervision for years. He made a trip for several months on the ocean, to fit him for work, but had collapsed again as soon as he began his eye-work in the office. Riding for a short time in a railway-car (if he read or looked out of the window) would give him indescribable distress in his head and compel him to go to bed. His family history shows the existence of phthisis, extreme neurasthenia, and severe headaches. A tendency toward melancholia in his ancestral lines has manifested itself in several instances. His brother has always suffered with headaches, and has been "weak-minded," so that he was unable to fill clerical positions with satisfaction to his employers for any length of time. Correction of a hyperphoria and a very high degree of esophoria by graduated tenotomies has effected in this patient a complete restoration to health.

Case XXX.—This lady was sent to me by her physician because he "could find nothing to relieve her." Her headaches, although extremely violent at times, never entirely subsided. Transient attacks of numbness of the face, lips, and limbs would occur at intervals. She showed both hyperphoria and esophoria. These were corrected by graduated tenotomies. Over ten years have now elapsed since the operation. Her headaches rarely occur, her physical strength has improved wonderfully, and she regards herself as practically cured.

Case XXXI.—This patient had been a source of anxiety to his parents from childhood on account of delicate health and nervous debility. Since manhood he had sought constantly for relief. "Nervous dyspepsia" had existed since early boyhood. Since the seventeenth year supra-orbital neuralgia has proved a source of almost constant distress. For over six months he was under the care of a prominent oculist, who ordered different reading-glasses from time to time, that failed to afford relief. He was then treated for nasal occlusion by chromic-acid applications, but no benefit was experienced. He had, at times, been under the medicinal treatment of many of the leading physicians, but had obtained no permanent relief from drugs, and had become very despondent in consequence. His mother and sister have neuralgia; the latter, of a very severe and intractable form. His father is very nervous, and suffered from asthenopia until I corrected his refractive errors. In a letter lately received by me from the patient he says: "Our benefactors are seldom remembered until a punching is received.

I have been feeling so well that I've not thought of how I got there, nor the cost, until your bill came."

CASE XXXIII.—This patient was a doctor. His hyper-metropia was totally latent, and was disclosed only after the use of atropine. The correction of his muscular and refractive anomalies caused a most favorable change.

CASE XXXIV.—This case is well worthy of record. The patient was a poor factory-girl that was sent to me by a medical friend in Massachusetts, to see if anything could be done for her, as no one who had seen her could make a diagnosis. Her appearance was one of remarkable deformity. Both upper limbs were contractured in the state of semiflexion at elbows, wrists, and fingers, and trembled excessively when she tried to use them. They could not be extended, nor could the patient make much use of the hands. She had been obliged to give up work on that account. When she talked, her face became painfully drawn and distorted. The mouth would especially be drawn downward. Under the least excitement she would be seized with what she called "choking-spells." The throat-muscles would contract and interfere seriously with respiration. She had constant headache—chiefly in the fore-head and occiput—and a persistent pain in the neck. No pain-ful points existed; nor did pressure, movement, or other tests reveal the existence of organic disease of the spinal cord. Sen-sibility to touch, pain, and temperature was normal. Motility was unimpaired. The muscular sense was perfect. No inco-ordination existed. Several photographs of this patient were taken at my request, but they failed to give an idea of the frightful contortions of the face and upper limbs that existed when the patient was unduly excited. I considered the case as one of aggravated chorea, complicated by headache and asthenopia.

An examination of the eyes showed a manifest hyperme-tropia ($+ 0.75$ D.) present, which remained unchanged under atropine. She also showed a slight esophoria (which under prisms became a homonymous diplopia) and a left hyperphoria of 3°. These muscular anomalies were satisfactorily relieved by several graduated tenotomies, and applications of static sparks were made daily for about two months. The patient has now regained the use of both hands, carries her head erect, has no headaches and little spinal pain, and for awhile returned to work. The effect of hard labor brought back her " choking-spells "; hence she was advised to take a year of absolute rest. She now has some asthenopia remaining, and an occasional headache. Her facial contortions persist to a slight extent when she is unduly excited. I suspect that some latent hyperphoria exists still, and that further operative work on the eye-muscles will be demanded before the patient recovers completely.

Case XXXVI.—This case illustrates the dependence of neurasthenia upon eye-strain. The symptoms which co-existed with headache entirely disappeared when the cause of excessive nervous expenditure was removed. A more pitiable nervous wreck than this patient was when sent to me is but seldom encountered.

Case XXXVII.—This was a simple case of eye-strain, due to a refractive and muscular error combined. The restoration to health seems to be complete. The headaches had become practically uninterrupted and had unfitted the patient for any occupation.

Case XXXVIII.—This case is one of peculiar interest because of the life-long duration of the symptoms and the various steps which had been taken to get relief. The patient was the wife of a physician, who had been obliged to abandon a lucrative practice in Chicago on account of his wife's health. She had been delicate from childhood, and eventually succumbed to constant headache, neuralgic paroxysms, failure of the digestive functions, and insomnia. She also developed tremor, hyperæsthesia of the entire body, emotional disturbances, and other symptoms of "nervous prostration." Her womb was supposed to be at fault, and all her other symptoms were deemed as echoes from pelvic disturbances. For fifteen years she had been under the best medical advice (chiefly gynæcological) in the various cities in which they had, at times, taken up their residence. The heredity was a bad one. Almost all of the ancestors and relatives (male and female), on both sides, had given evidence of nervous troubles. The menopause had been confidently waited for as the millennium in which the supposed ovarian disturbance would cease; but the hope had proved a delusive one, as no improvement seemed to come.

There had never been, until after the menopause, any trouble with the eyes of which the patient was conscious, in spite of the fact that the cause of her nervous disturbances had, without doubt, been chiefly ocular throughout her life of suffering. She had reached a point, before I saw her, where seclusion from the society of friends and confinement to her own room had become a necessity. The headaches were intolerable, the use of the eyes a source of discomfort, and the physical weakness extreme. When I first saw her she was brought to my office in a carriage. Within three weeks two graduated tenotomies had been performed upon her eye-muscles. She attended a place of amusement soon after the first operation, which gave her great relief. In less than two months she undertook the care of a home of her own in Brooklyn, and visited me daily at my office for some time. She has had but little headache since the operative treatment upon her eye-muscles was begun. She still suffers, however, from acute digestive disturbances, at times, which retard her recovery. At the last examination I discovered some hyperphoria which had not been relieved. This factor may be an active one in causing her gastric derangement.

CASE XXXIX.—Although, in this case, a complete cessation of headaches for six months followed the use of prismatic glasses, the parents refused to allow the use of atropine and a radical correction of the hyperphoria and esophoria. As the patient lives at a distance from New York, I have no record of her present muscular tests.

CASE XL.—This patient had one eye that was too defective to be of much use. Her tests were very unsatisfactory in consequence, and little encouragement was given her. Her ocular conditions were unquestionably a very important factor in her nervous debility.

CASE XLI.—This patient was a divine and a victim to chronic sick headaches, asthenopia, insomnia, and general neurasthenia. He had been obliged to desist from pastoral work in consequence of his many sufferings. At the last interview he told me that he had recovered so rapidly that he had resumed pastoral work, and now suffered but little from his old symptoms. Too close application in his study tends to cause him a slight return of distress in his head and eyes. He no longer has any double vision.

CASE XLIII.—This case was a long and tedious one, and the results are, perhaps, all I am justified in expecting. I had to combat a high degree of astigmatism for the right eye, unequal refraction in both eyes, crossed diplopia (transient), and left hyperphoria of a very high degree (mostly latent). In spite of the difficulties of the case and the amount of operative work done upon his eyes, this patient is one of my warmest and most enthusiastic advocates. He certainly has been greatly relieved of most excruciating attacks of headache, which were becoming almost incessant. His asthenopia is also greatly improved. He is still under observation, with a hope of greater benefits.

CASE XLIV.—This case was one of unusual interest to me. The patient had succumbed to headache, insomnia, etc., and had had an attack of the so-called "nervous prostration," which compelled him to abandon all business pursuits and take a trip to Europe, which he could ill afford. This trip gave no permanent benefit, and melancholia with suicidal tendencies developed after his return. No medicinal treatment yielded any results. He was then referred to me by his brother (a physician of prominence in this city). Graduated tenotomies were performed for esophoria and hyperphoria. The effect was immediate and startling to his family. He completely regained his health, strength, and mental vigor within six weeks. Over ten years have elapsed without any signs of return of his old symptoms. Moreover, he has since passed through a serious financial disaster with a buoyancy that surprised himself and others.

CASE XLVI.—This case illustrates, in a remarkable way, a high degree of latent hypermetropia. Spherical glasses, in this case, have entirely arrested the headaches as well as convulsive attacks (to which the patient was subject) for two years.

CASE XLVII.—This patient was a physician. He reports that very marked improvement—not only in his headaches and asthenopia, but in his general health—has followed the relief of his eye-strain.

CASE L.—The final case reported is one of the most remarkable that I have ever encountered. The patient was brought to my office, from her home in Virginia, by her sister, who wrote to me as follows, before her arrival:—

"My sister at one time thought her headaches came from her teeth, and she had twenty-six drawn. This may give you some idea how tremendously in earnest she is about being cured."

When a child, headaches would appear after a transient attack of blindness. They increased in frequency, so that, after marriage, two a week often occurred (each lasting twenty-four hours and followed by two days of confinement in bed from exhaustion). When pregnant she had several "falling fits" daily, and remained totally unconscious for some time. She seldom passed a week without being unconscious or semi-comatose from pain. The slightest exertion caused "palpitation of the heart" to such an extent as to cause a fear of approaching death. A correction of the hyperphoria by graduated tenotomies caused almost complete restoration of health.

The patient reports that, during the past six months, "she has had but three headaches,—two light and one severe, the latter being caused by intense emotional excitement." I learn from her, also, that she can bear fatigue without any of the bad symptoms that used to inevitably follow. She was only under my observation a very short time, hence I suspect that the hyperphoria may yet be imperfectly corrected. Before she left New York she journeyed some fifteen miles daily to visit my office, and experienced no distress after the effort.

The tabulated record of these fifty cases may be particularly interesting to the reader because the cases were not specially selected from my case-books, but are consecutive.

I know of no similar table in medical literature. No drugs were employed during the treatment of any of these patients, in order that the results could not be attributed to any other possible agent than the eye-treatment.

Almost every case reported here was a chronic case, where the headache or neuralgic paroxysms had never been benefited for any length of time by medicinal or dietetic treatment.

Case.	Age.	Sex.	Condition.	Symptoms.	Duration.	Heredity.	Refraction.	Muscular.	Treatment.	Remarks on Case.
1	50	M.	Married.	Sick headaches (very frequent and severe). Asthenopia.	Since early childhood.	Headaches in all paternal ancestry and in brothers and sisters. Phthisis common among maternal ancestry.	R. +0.75 s. L. +1.00 s. +0.50 c.	Right hyperphoria, 2½°. Exophoria, 10.	Full correction of refractive error by glasses. 10 prism for R. hyperphoria.	In this patient latent hyperphoria is suspected to exist. No headaches have occurred since the glasses have been worn. The muscular adjustment of the eyes should be rectified by a graduated tenotomy.
2	25	M.	Single.	Sick headaches. A peculiar nervous hesitancy of speech. Mental depression of a marked character. Mental weakness. Extreme nervousness and excitability.	Since 12th year. do. do. Since childhood. Since 12th year.	Father and two brothers have severe headaches. Ancestry are peculiarly nervous and excitable.	Manifest emmetropia. Atropine not used.	Emphoria, 150. Homonymous dipl pin, 30 (with red glass).	Graduated tenotomy of both internal.	Patient had to leave for the West. He refused atropine tests on that account. After the headaches the power of abduction was increased from 10 to 80. No headaches have occurred since operation. Patient lately reports a great change in his general health and a sense of rest in his eyes never experienced prior to the operation. The mental vigor of this patient already shows a marked improvement.
3	20	F.	Single.	Supra-orbital neuralgia. Frequent fainting attacks. Complete amenorrhœa. Melancholia. Dyspepsia.	Since child-hood. do. For 2 years. Since child-hood.	Mother had frequent fainting attacks. One brother has severe headaches. Two cases of epilepsy on paternal side. Phthisis very common on maternal side.	R. +0.50 s. L. +0.50 c. +1.25 c.	Right hyperphoria, 6°. Exophoria, 10.	50 R. hyperphorial prism given.	Patient has been almost entirely relieved of her neuralgic paroxysms by the glasses ordered. She is much less depressed and has had no fainting attacks. A graduated tenotomy should be performed to correct the hyperphoria.
4	30	M.	Married.	Asthenopia. Occipital headache. Insomnia. Hesitancy in speech. Dyspepsia. Asthenopia.	For 2 years. Since child-hood. do. do.	Father is a dyspeptic and rheumatic. Mother is at times insane. Headaches are a family trait. One uncle was insane. One cousin had epilepsy.	R. +1.50 s. L. +1.25 s.	Right hyperphoria, 20.	Patient was given for constant wear: R. + 1.50 s. L. + 1.25 s.	This patient's attacks were often accompanied by symptoms of petit mal. He has experienced great relief from his headaches since his glasses were ordered. No attacks of lune of consciousness have occurred since the glasses were prescribed.
5	53	M.	Married.	Severe headache, frontal and occipital. Insomnia of an intractable type. Prostatic and urethral irritation, causing the patient great anxiety and discomfort.	For 2 years. Since child-hood. For 2 years. For 2 years.	Father had severe headache. Mother and two sisters had phthisis.	R. +1.50 s. L. +1.50 s. +0.50 c.	Right hyperphoria, 1°. Exophoria, 10.	Patient was given + 1.50 s. + 1.50 prism to each eye to correct right hyperphoria.	This patient expresses himself as almost completely cured of headache and insomnia by his glasses alone. His intense eye-strain has thus far prevented neuralgia and has been decidedly toward the correction of his hyperphoria. His prostatic neuralgia is very decidedly relieved without any hard treatment thus far.

* This report is guarded about as it was originally published in July, 1888. Some of those cases have since been lost sight of and their addresses to-day are irrecoverable to the author; others have been seen or heard from and their condition has been noted in this table. Subsequent chapters contain many histories where headache and neuralgia were relieved by eye-treatment, as well as more serious nervous disturbances.

A TABULATED RECORD OF FIFTY CONSECUTIVE CASES OF HEADACHE AND NEURALGIA TREATED EXCLUSIVELY BY THE CORRECTION OF EYE-STRAIN (continued).

Case.	Age.	Sex.	Condition.	Symptoms.	Duration.	Heredity.	Refraction.	Muscular.	Treatment.	Remarks on Case.
6	5	F.	Married.	Neuralgia (very severe and chiefly of the supra-orbital and occipital types).	Since 5th year.	Father is very nervous. Mother has neuralgic headaches. Paralysis in paternal ancestry. Phthisis on maternal side.	R. + 0.75. L. + 0.75.	Left hyperphoria, 6°. Esophoria, 0 to 6°.	Graduated tenotomy of right inferior rectus.	This patient reports that a great change for the better has occurred since the operation. Her neuralgic attacks have been decreased 75 per cent, and her physical condition is greatly improved.
7	13	M.	Single.	Headache (in frontal and occipital regions). Has been almost constant and is often accompanied by vomiting.	3 years.	Headaches and eye-defects are common on the maternal side, (One paternal uncle has headaches, and another incipient ataxia. Rheumatism is a family trait on paternal side.	R. + 0.25. L. + 0.50 c. L. + 0.50 c.	Left hyperphoria, 3°.	Given for distance, +0.50 c. Graduated tenotomy of right inferior rectus.	This patient has had no headache for several months since the operation. He still suffers from occasional gastric disturbances. There is reason to suspect some latent hyperphoria in this case, which may demand further operative work.
8	13	F.	Married.	Sick headaches (so frequent and severe as to unfit the patient for household duties for the past ten years).	Since childhood.		R. + 1.00 c. L. + 1.00 c.	Right hyperphoria, 5°. Esophoria, 25 to 85°.	Given for constant wear, +1.00 c. Graduated tenotomy of left inferior rectus.	This patient reports a very marked decrease of her headaches since the operation, and a remarkable increase of physical strength. The had reported before that she "can endure fatigue better than for many years and has headaches only at long intervals." Further operative work will probably be demanded in this case.
9	35	F.	Single.	Supra-orbital and occipital neuralgia. Asthenopia (of many years' standing). Dyspepsia and vomiting.	Since 12th year. For 6 years past. For many years.		R. 2.25 c. L. 1.00 c.	Right hyperphoria, 2°. Esophoria, 10°.	Given for constant wear, O. D. +2.25 c. O. S. −1.00 c. Graduated tenotomy of left inferior rectus and of both internal recti.	This case has proved to be a very difficult one to treat. She had homonymous diplopia, together with a marked hyperphoria. Her ability to use the eyes in sewing and reading has been greatly improved, and her neuralgic paroxysms and vomiting have been markedly benefited, but not arrested as yet. Further operative work should be done later on.
10	30	F.	Married.	Frequent and severe sick headaches. Asthenopia (of a severe type). Nervous prostration (lasting for many months).	Since 9th year. Several years. 5 years ago.	Father and sister are subject to sick headaches. Maternal grandmother has neuralgia.	R. + 1.25 s. L. + 1.25 s.	Right hyperphoria, 5°. Esophoria, 15°.	Graduated tenotomy of left inferior rectus and of both interni.	The last reports from this patient give assurances of almost a perfect cure. Her headaches have disappeared. Her asthenopic symptoms are subsiding fast. Her physical strength is almost entirely restored.
11	11	M.	Single.	Frequent and protracted headaches—had most of the time—are frontal and occipital. Asthenopia. Obstinate constipation.	2 years. 4 years. Since childhood.	Phthisis and epilepsy in both paternal and maternal ancestry. One brother has epilepsy.	R. + 0.75 s. L. + 0.50 s. L. + 0.25 c.	Esophoria, 12°. Left hyperphoria, 3°.	Graduated tenotomy of left internal and right inferior rectus.	This case requires further observation and has proved to be quite intractable. The muscular anomalies of the orbit have been extreme, and sufficient time has not elapsed for the patient to get the full benefits of their correction. He reports that he has "much less asthenopia and headache, and is now able to study with little distress."

CASE	AGE	SEX	CONDITION	SYMPTOMS	DURATION	HEREDITY	Refraction	Muscular	TREATMENT	REMARKS ON CASE
12	30	M.	Single.	Headaches (chiefly frontal). Melancholia. Constipation.	18 months.	Mother died of cancer; father of typhoid fever; maternal aunt died of phthisis. Headaches very common on both sides.	R. $+0.25$. -1.00 c. L. Emmetropia.	Esophoria, 7°.	Given for constant wear: $+0.25$ () -1.00 c. for right eye. Graduated tenotomy of both internal.	This patient had been under the care of many prominent physicians without obtaining any relief from drugs. He writes me as follows: "I pass many pleasanter days now than for many years past." His improvement, when last seen by me, was very apparent. His headaches are less frequent and severe.
13	25	F.	Married.	Sick headaches (very severe and frequent). Asthenopia.	Since childhood.	Father died of heart disease. Mother had severe headaches and vertigo. Maternal grandfather was insane.	R. $+0.50$ s. L. $+0.50$.	Right hyperphoria, 1°. Esophoria, 20.	Graduated tenotomy of left inferior rectus. 10 prism for left hyperphoria still remaining.	This patient had been under constant medical care for years before I saw her. Her symptoms were generally attributed to uterine trouble, but all treatment so directed proved unavailing. Although her eye-muscles are not yet perfectly adjusted, she has experienced a very marked change for the better. Her headaches and asthenopia have been decreased over 75 per cent., and the pain in the back has almost entirely ceased.
				Weakness in arms and legs, also fainting attacks. Severe pain between the shoulder-blades.	For past 3 years, do. do.					
11	46	M.	Married.	Sick headaches. Nervous prostration. Nervous weakness and morbid fears after any unusual strain.	Since childhood. 9 years ago. Several years.	Father had severe headaches and died of Bright's disease. Mother died of cerebral softening. One brother died of "neuralgia of brain." Mental diseases on paternal side are common.	R. $+1.00$. $+0.25$ c. L. $+1.00$. $+0.25$ c.	Esophoria, 6°.	Given for constant wear: $+0.25$ c. $+0.25$ c. Graduated tenotomy of left internal rectus.	This patient still has an occasional headache, but at long intervals. He expresses himself, in a late communication, as delighted with the marked change which has occurred. His physical condition has improved rapidly since the operation.
15	28	F.	Single.	Constant headache. Pain in the back, which has proved very rebellious. Asthenopia. Physical weakness (to such an extent as to render walking almost impossible). Insomnia.	3 years. Since 11th year. Since childhood. 10 years.	Headaches and neurasthenia very common on maternal side.	R. $+0.50$. L. $+0.50$.	Esophoria, 15°. Homonymous diplopia of 20 (with red glass). Right hyperphoria, 4°.	Graduated tenotomies of both internal, and of the left inferior and right superior recti.	This patient wore a plaster jacket for over a year for her spinal pain; and was also treated for uterine conditions, without any relief to her symptoms. Her ocular conditions proved to be very complicated and difficult to fully correct. She now is almost entirely free from spinal pain, has only occasional headaches, and is able to walk several miles each day.
16	16	F.	Single.	Constant headache (without any intermission). Chorea, which had practically disappeared before I saw the patient.	Since 8th year. 1 in childhood.	Father has had several attacks of petit mal. Mother had sick headaches and nervous prostration. Mother's family is very nervous. Paternal grandmother and much died of phthisis.	R. $+1.00$ s. L. $+1.00$ s.	Left hyperphoria, 7.4°, most of which was latent. Exophoria, 9°.	Graduated tenotomies of right inferior rectus, left superior rectus, and both externi.	A marked improvement in this case has been effected, but the patient still has headaches at times, which is always intensified by close application of the eyes.

A TABULATED RECORD OF FIFTY CONSECUTIVE CASES OF HEADACHE AND NEURALGIA TREATED EXCLUSIVELY BY THE CORRECTION OF EYE-STRAIN (continued).

Case.	Age.	Sex.	Condition.	Symptoms.	Duration.	Heredity.	Refraction.	Muscular.	Treatment.	Remarks on Case.
17	36	F.	Married.	Headaches, always after any fatigue, as shopping. Asthenopia.	For some years.	Mother died of phthisis. One sister is a victim to headaches.	R. + 1.25 c. L. + 1.25 c.	Exophoria, 30°.	Given for constant wear: + 1.00 c.	This patient made a complete recovery by the aid of cylindrical glasses. The last report received was that she "had been for several months absolutely free from headache."
18	38	M.	Married.	Sick headaches (these headaches lasted 2 or 3 days in each week). Indigestion. Constipation.	5 years.	Father died of gastric trouble. One brother has headaches.	R. + 1.50 s. + 0.25 c. L. + 1.50 s.	Esophoria, 80°.	Given for constant wear: + 2.50 s. Graduated tenotomy of right internal rectus.	Patient left for his home in Kansas City, Mo., with instructions to report the results of treatment. Although many months have elapsed, no report has been received.
19	51	F.	Married.	Supra-orbital neuralgia. Insomnia. Marked twitching of the eyelid.	12 years. do.	One sister has neuralgia.	R. + 0.75 c. + 0.25 c. L. + 1.00 c. + 0.50 c. Reads No. 1 Jaeger's type with + 1.00 s.	Esophoria, 20°.	Given for constant wear: O. D. + 0.75 c. O. S. + 0.50 c. For reading + 1.00 s. was added to each of the cylinders.	This patient has reported that she is "completely cured and is in perfect health." She has had no pain for over a year.
20	55	F.	Single.	Sick headaches. Asthenopia. Constipation.	Since 15th year. 1 year. do.	Mother had headaches when young. Phthisis on maternal side.	R. + 1.50 s. L. − 0.50 s.	Esophoria, 60°.	Given for distance: − 0.50 s.	This patient refused surgical treatment of her exophoria. She would not tolerate prisms, as they caused distress. She returned to the West without any benefit.
21	11	F.	Married.	Neuralgia. Nervous prostration (three attacks). Chronic constipation.	20 years. do.	One brother and one sister died of phthisis. Phthisis common on paternal side.	R. + 1.75 s. L. + 1.25 s.	Esophoria, 30°.	Given for constant wear: + 1.00 s.; for reading + 1.50 s. Graduated tenotomy of left internal rectus.	The last report from an oculist to whom I intrusted this case, after her departure from New York, gives assurances of a decided physical improvement, but not, as yet, a radical cure of the neuralgia. A latent hyperphoria is suspected to exist.
22	47	M.	Single.	Headaches (very severe, occurring 3 or 4 times a week).	Since 15th year.	Mother died of cancer.	R. 2.00 s. L. − 2.00 s.	Esophoria, 40°.	Ordered for constant wear: − 1.50 s., combined with 1° prism, base out, in each eye.	This patient had gone over six months without a headache when last heard from. He expresses himself as being "astonished" at the magical change which followed the use of glasses.
23	33	F.	Single.	Sick headaches (very severe and frequent). Morbid and uncontrollable fear of company, sudden noises, etc. Very emotional. Twitching of the eye-ball. Difficulty in raising the upper lids.	Since early childhood. 4 years. do. do. do.	Father died of phthisis. Phthisis common in paternal ancestry. One maternal uncle died of phthisis.	R. + 0.50 s. L. + 0.50 s. + 1.00 s.	Exophoria, 5°. Left hyperphoria, 40.	Given for constant wear: O. D. + 0.50 s. () + 0.50 s. O. S. + 1.00 s. Graduated tenotomies of both external and right inferior recti.	This case was one of great interest to me. She had been under the care of several oculists, who had simply corrected the astigmatism of the right eye. Her mental condition caused serious alarm. The results of treatment have been very gratifying. Patient has entirely recovered control of her emotions, and has but few headaches. She often passes many consecutive weeks without any pain.

A TABULATED RECORD OF FIFTY CONSECUTIVE CASES OF HEADACHE AND NEURALGIA TREATED EXCLUSIVELY BY THE CORRECTION OF EYE-STRAIN (continued).

CASE	AGE	SEX	CONDITION	SYMPTOMS	DURATION	HEREDITY	VISUAL TESTS — Refraction	VISUAL TESTS — Muscular	TREATMENT	REMARKS ON CASE
21	45	M.	Married.	Neuralgic headaches (very severe and almost constant). Vertigo (after looking intensely at any object). Dyspepsia and acid stomach. Confusion of thought. Insomnia.	Since early childhood.	Father had headaches. Mother died of brain disease. Every one of his maternal ancestry suffers severely with headache and neuralgia. One brother died of brain disease. Two sisters suffer terribly with headache.	Emmetropia. Presbyopia. + 1.25 s.	Esophoria, 12°. Left hyperphoria, 10°.	Graduated tenotomies of both internal and right inferior rectus.	Patient was a physician. He had regarded his condition as incurable. A remarkable heredity existed as regards headache, neuralgia, dyspepsia, and brain diseases. He is now practically well. An occasional headache and a weak stomach indicate that the eye-strain is not entirely removed. He still shows some emphoria at times.
22	31	M.	Married.	Neuralgic headaches (very severe and only controlled by large doses of morphine). Habitual constipation. 3 per cent. of sugar in the urine (analysis of Dr. G. T. Elliott, of New York).	6 years.	Mother had severe neuralgic headaches.	R. + 1.50 s. L. + 1.50 s.	Esophoria, 5°. Left hyperphoria, 3°.	Given for constant wear: + 0.50 s. Graduated tenotomies of right internus and left superior rectus.	This case was pronounced by Dr. G. T. Elliott, of New York, as "one of typical diabetic headache." Without any drugs his sugar was reduced to a half of one per cent. within a few weeks, by the relief of abnormal muscular tension in the orbit. The neuralgic attacks were totally arrested for a time, but the patient has thus far failed to allow of a complete restoration of muscular equilibrium.
23	29	F.	Single.	Severe sick headaches. Asthenopia.	Since childhood. do.	Father suffers with neuralgia, two brother suffers with severe headaches.	R. + 3.00 s. L. + 2.50 s.	Esophoria, 6°.	Given for constant wear: O. D. + 1.50 s. O. S. + 1.00 s. Graduated tenotomies of both externi.	Very marked benefit, for a time, in regard to headaches and asthenopia. Further correction of a latent esophoria that disclosed itself after the tenotomies were advised, but the patient has thus far failed to report at my office.
24	24	M.	Single.	Constant dull headache. Habitual constipation. Cold hands and feet. Difficulty in concentrating the mind.		Father and several paternal relatives died of phthisis. Mother has nephritis and has had left hemiplegia. One sister has hystero-epilepsy and asthenopia.	R. + 1.00 s. L. + 1.00 s.	Esophoria, 10°. Right hyperphoria, 3¼°.	Given for constant wear: + 1.00 s. Graduated tenotomies of the right interni and the right superior rectus.	In this case the results were at first slow, in spite of patient watching and repeated eye-tests. The patient has also been treated for disfiction of the nasal septum by Dr. Bokavan, of New York, without relief to his head. Professor J. W. Wright, of New York, also, at my suggestion, treated the patient for an anal fissure and hæmorrhoids. A marked improvement of headache has followed during the past five years. He is now cured.
25	24	M.	Married.	Sick headaches. Insomnia. Mental depression. Sense of pressure in occipital region.	4 years. do. do.	Headaches are common in the maternal ancestry.	R. + 0.50 s. + 0.75 c. L. + 1.00 c.	Esophoria, 20°.	Given for constant wear: O. D. + 0.25 c. O. S. + 10 prism, base out.	Patient has never reported the results of wearing his glasses. Friends of the patient state that he is now able to attend to his business in a distant city much more regularly than before.

A TABULATED RECORD OF FIFTY CONSECUTIVE CASES OF HEADACHE AND NEURALGIA TREATED EXCLUSIVELY BY THE CORRECTION OF EYE-STRAIN (continued).

CASE.	AGE.	SEX.	CONDITION.	SYMPTOMS.	DURATION.	HEREDITY.	Refraction.	Muscular.	TREATMENT.	REMARKS ON CASE.
29	27	M.	Single.	Neuralgic headaches. Intractable insomnia. Asthenopia. Indigestion. General nervousness.	Since childhood.	Father is very nervous and has severe headaches. Two brothers are also very nervous and have severe headaches.	Emmetropia.	Left hyperphoria, 2°. Esophoria, 15°.	Graduated tenotomies of right inferior rectus and of both interni.	This patient is now entirely cured of headaches and his nervous dyspepsia. His asthenopia has almost entirely disappeared.
30	25	F.	Married.	Sick headaches. Difficulty in raising the eyelids. Transient attacks of numbness of lips and paresis of limbs. Constant headache.	Since childhood. 3 years. do.	Father died of an abscess. One aunt has severe headaches.	R. + 0.50 s. L. + 0.50 s.	Left hyperphoria, 1½°. Esophoria, 12°.	Graduated tenotomies of right inferior rectus and of right internal rectus.	Patient reports her condition as one of practical restoration to health. Her headaches are rarely developed, and only after some overexertion, loss of sleep, or fatigue.
31	27	M.	Single.	Supra-orbital neuralgia (almost without cessation). Dyspepsia. Pain in right eye. Has been in very delicate health.	10 years. do. do. Since childhood.	Father suffers with his eyes. Mother and sister have neuralgia. One maternal uncle died of phthisis.	R. + 1.00 s. L. + 1.00 s.	Esophoria (totally latent) amounting to 17°.	Graduated tenotomies on both interni.	This patient has shown some esophoria that was not manifested for many months after the last operation. He has been restored to health and comfort, and has passed many months without pain.
32	43	M.	Married.	Sick headaches. Dyspepsia. Severe insomnia. Partial right hemianæsthesia. Transient diplopia.	Since childhood. 25 years. 5 years. 6 years ago.	Father had headaches. Two sisters and two cousins died of phthisis.	R. + 1.00 s. L. + 1.00 s.	Esophoria, 30° (mostly latent). Left hyperphoria, 2°.	Graduated tenotomies of both interni and of right inferior and left superior.	Headaches were improved, but no efforts upon the distressing insomnia were obtained. The patient left for the distant West, to live on a ranch, before his esophoria was fully corrected.
33	36	M.	Married.	Sick headaches. Nervousness and irritability. Weak heart. Indigestion.	When a child. Since childhood.	Phthisis on maternal side. Paternal ancestry all dyspeptics. Maternal grandfather insane for a time. One maternal uncle "queer," disappeared suddenly, and not heard from. Patient knew little of her ancestry.	R. + 2.25 s. L. + 2.00 s.	Esophoria, 13°.	Given for constant wear: O. D. + 1.75 s. O. S. + 1.00 s. Graduated tenotomies of both interni.	This patient reports that he is "practically well of headaches, and cannot get along without his glasses." He has gone to Kansas City to practice medicine.
34	36	F.	Single.	Neuralgic headaches. Asthenopia. Persistent trembling in hands and arms. Rigidity of neck and choking spells when attempting to speak or when excited. Severe pain in neck and arms. Contracture of hands and arms.	Since 10th year. do. 4 years. do. do.		R. + 0.75 s. L. + 0.75 s.	Esophoria, 20° (mostly latent). Left hyperphoria, 3°.	Graduated tenotomies of both interni and of right inferior and left superior recti. A full correction of hypermetropia by glasses.	For full report, see history of this case. The patient is practically cured, but has some chronic symptoms remaining, which appear under undue excitement.

A TABULATED RECORD OF FIFTY CONSECUTIVE CASES OF HEADACHE AND NEURALGIA TREATED EXCLUSIVELY BY THE CORRECTION OF EYE-STRAIN (continued).

CASE.	AGE.	SEX.	CONDITION.	SYMPTOMS.	DURATION.	HEREDITY.	VISUAL TESTS. Refraction.	Muscular.	TREATMENT.	REMARKS ON CASE.
35	25	M.	Single.	Sick headaches. Nervousness and some tremor when excited. Partial impotence.	Since childhood. do.	Mother has headaches. One brother and one sister have severe headaches.	R. +0.50 s. L. +0.50 c.	Esophoria, 11°.	Graduated tenotomies of both interni.	Cured of headache. The patient is much less nervous, and has nearly regained his sexual powers.
36	25	M.	Single.	Headaches. Periodical attacks of vomiting and indigestion. Slow heart's action. Cold extremities. Physical weakness. Irregular action of bowels (constipation and diarrhœa).	5 years. Since childhood. do. do. do. do.	Mother has headaches, and has had pulmonary haemorrhages. Father has neuralgia of the stomach. Maternal grandmother and two uncles died of phthisis.	R. +0.50 s. L. +0.75 c.	Esophoria, 20° (mostly latent).	Graduated tenotomies of both interni. Spherical glasses for reading.	This patient was a divinity student. He was sent to me from Vermont. His recovery was rapid and complete. His last communication is full of expressions of delight over his ability to pursue his studies.
37		F.	Single.	Headache (violent sick headache, which during past few years became almost uninterrupted). Obstinate dyspepsia.	Since childhood.	Father suffered with headache and died of hemiplegia. Phthisis in maternal ancestry.	R. +3.00 s. L. +3.00 s.	Esophoria, 9°.	Given for constant wear: +1.75 s. Graduated tenotomies of both externi.	Patient reports that she has "not had a headache for past six months"; also that her "physical condition is better than at any time in her life."
38	52	F.	Married.	Sick headaches (very severe and frequent). Attacks of tremor. Hyperaesthesia of entire body. Neuralgic attacks. Insomnia.	Since childhood. do. 30 years. do. do.	Father had severe headaches and was slightly crosseyed. Mother was a confirmed invalid. One sister has terrible headaches. One brother is a great nervous sufferer. Headaches and nervous prostration are family traits on both sides.	R. +1.25 s. L. +1.25 s. +0.25 c.	Esophoria, 11° (mostly latent). Right hyperphoria, 1/2°.	Given for constant wear: +1.25 s. O. D. +1.25 s. +0.25 c. O. S. Graduated tenotomies of both interni and of the left inferior rectus.	Headaches almost totally arrested. Insomnia cured, as is also the tremor and hyperæsthesia.
39	21	F.	Single.	Sick headaches (very severe and constant when she uses her eyes). Asthenopia.	Since 15th year. do.	Father, mother, and sister have headaches and neuralgia and weak eyes. Phthisis frequent among paternal ancestry.	Manifest emmetropia. Atropine not used.	Esophoria, 10°. Right hyperphoria, 2°.	Given for constant wear: 2° prism, base out, in each eye.	This patient went nearly six months without a headache after her prisms were ordered. The parents refused atropine and operative procedure. She has had some return of headache lately.
40	46	F.	Single.	Headaches (chiefly orbital and frontal). Neuralgia in fifth nerve. Obstinate insomnia. Constipation. Palpitation of heart.	Since childhood.	Mother, maternal grandmother, one brother, and one sister had phthisis; also several paternal aunts. Father died of chronic diarrhœa.	R. +3.50 c. L. +1.50 s. Presbyopia, +3.00 s.	Esophoria, 6°.	Given for constant wear: +1.50 c. O. D. +1.50 s. O. S.	Results negative. The patient had one very defective eye which could not be rectified, and which made her tests very unreliable. No encouragement of marked improvement could be given her.

A TABULATED RECORD OF FIFTY CONSECUTIVE CASES OF HEADACHE AND NEURALGIA TREATED EXCLUSIVELY BY THE CORRECTION OF EYE-STRAIN (continued).

Case.	Age.	Sex.	Condition.	Symptoms.	Duration.	Heredity.	Refraction.	Muscular.	Treatment.	Remarks on Case.
41	38	M.	Married.	Sick headaches. Intense suffering from light. Tight band at temples and numbness of head. Insomnia. Irritability and physical weakness. Diplopia.	Since childhood. 2 years. do. do.	Father died of heart disease and was very nervous. One sister has severe headaches. Most of the family have severe migraine. One paternal cousin died of phthisis.	Emmetropia.	Esophoria, 9°. Right hyperphoria, 3°.	Graduated tenotomies of both interni and of left inferior rectus.	The patient made a practical recovery. He has headache and asthenopia only at intervals, and always from excessive use of the eyes in preparing sermons, etc.
42	25	M.	Single.	Frontal headache. Neuralgia. Asthenopia. Intolerance of light.	At times. 5 years. Until 1887. Since childhood. do.	Mother had headache and neuralgia and died of a tumor. One brother has neuralgia. Insanity and hypochondria common on maternal side. Maternal grandfather had cancer. Phthisis common on paternal side; also rheumatism.	R. + 1.00 s. L. + 1.00 s.	Esophoria, 7°.	Given for constant wear: $\bigcirc$ + 1.20 s.; $\bigcirc$ prism, base out, in each eye.	Patient reports that he now practices medicine with greater comfort than for many years. He has little, if any, asthenopia or headache.
43	58	M.		Sick headaches (very violent). Asthenopia.	Since youth. do.	Phthisis common on paternal side; also rheumatism.	R. + 2.25 s. + 1.50 c. L. + 1.25 s. + 0.50 c.	Esophoria, 19°. Left hyperphoria, 1°.	Given for constant wear: $\bigcirc$ + 2.25 s., $\bigcirc$ + 1.50 s.; $\bigcirc$ + 1.25 s., $\bigcirc$ + 0.50 c. Graduated tenotomies of both externi and right inferior rectus.	This patient's age was against him. His refractive conditions were unlike and hard to perfectly correct. He reports a very marked decrease of headache and asthenopia. He is still under my observation.
44	41	M.	Married.	Headache. Insomnia. Morbid feelings with delusions. Melancholia with suicidal tendency.	Since 20th year. 2 years. do. do.	Father died of tumor. Mother is a chronic invalid. Sister has slight hysteria. Paternal grandmother died of phthisis. One paternal uncle a suicide. One aunt was insane.	R. − 3.50 s. 1.00 c. L. − 3.50 s. 1.00 c.	Esophoria, 2°. Right hyperphoria, 2°.	Given for constant wear: − 3.50 s. C − 1.00 c. − 2.50 s. C − 1.00 c. Graduated tenotomies of both external and left inferior recti.	Complete restoration to health. No return of symptoms for past two years.
45	40	F.	Married.	Neuralgia. Sense of pressure in the head. Vertigo. Morbid fears. Insomnia.	Several yrs. do. do. do.	Two second-cousins insane.	R. + 1.00 s. L. + 1.00 s.	Esophoria, 10°.	Graduated tenotomies of both interni.	Patient promised to return for further observation, but has not yet done so. The last report was one of marked improvement.

A TABULATED RECORD OF FIFTY CONSECUTIVE CASES OF HEADACHE AND NEURALGIA TREATED EXCLUSIVELY BY THE CORRECTION OF EYE-STRAIN (concluded).

CASE	AGE	SEX	CONDITION	SYMPTOMS	DURATION	HEREDITY	Refraction	Muscular	TREATMENT	REMARKS ON CASE
46	19	F.	Single.	Headaches (severe and constant). Convulsions of the epileptic type. "Fainting attacks." Nervous prostration.	3 months. When child. During past 6 months. 6 mos. ago.	Mother is very nervous.	R. + 5.50 s. L. + 3.00 s.	Exophoria, 8°.	Given for constant wear: O. D. + 3.50 s. O. S. + 1.50 s.	Patient has worn the glasses for two years without a headache or convulsive seizure.
47	47	M.	Married.	Sick headaches. Watering of the eyes in cold weather. Asthenopia.	Since 16th year. 3 years. do.	Mother died of schirrus; father of prostatic trouble. Brother died of phthisis.	R. + 0.75 s. L. + 0.50 s.	Right hyperphoria, 1°. Exophoria, 1°.	Graduated tenotomy of right superior rectus.	Patient left for his home in California. He reports a very decided relief from headache and asthenopia.
48	34	F.	Married.	Sick headaches. Asthenopia. Constipation.	Since 8th year. 10 years. do.	Mother, maternal grandmother, one maternal aunt, one maternal uncle, one brother, and one sister died of phthisis.	R. + 3.00 s. L. + 3.00 s.	Esophoria, 15°. Left hyperphoria, 1°.	Given for constant wear: + 2.00 s.	Patient has been benefited by glasses ordered. Her husband refused consent to an operation for the muscular maladjustment.
49	32	M.	Single.	Headaches.	Since childhood.	Father has severe headaches. Mother died of puerperal fever. One brother has headaches.	R. - 0.50 c. L. - 0.25 s.	Esophoria, 8°. Left hyperphoria.	Given for constant wear: O.D. 10 prism, base up. O. S. 30 prism, base out.	No report has ever been received from this patient. He lives in the West and has not returned for further treatment.
50	19	F.	Married.	Neuralgic headaches (causing coma of 21 hours' duration).	Since 11th year.	Father died of paralysis. Mother had "suffocation spells." One sister is very nervous. Dyspepsia is a family trait.	R. - 2.50 s. + 0.75 c. L. - 2.50 s. + 1.50 c.	Left hyperphoria, 1°.	Given for constant wear: - 2.50 s. ○ O. D. + 0.75 c. - 2.50 ○ O. S. + 1.50 c. Graduated tenotomies of left superior and right inferior recti.	Total cessation of headache for a long period of time. (Occasional headaches according to last report) after severe exertion or excitement.

CHAPTER IV.

THE EYE-TREATMENT OF CHOREA.

(St. Vitus's Dance.)

A Critical Review of Certain Factors that May Lead to Spasmodic Diseases and the Treatment of Such Conditions Without Drugs.

My attention has been attracted by a most remarkable statement that has lately appeared,—viz., the following quotation from a *brochure*, by an author of repute,[1] that has been published relative to the causation of chorea. He says: "The more recent studies of the pathology of chorea have led to a practically *unanimous*[2] conclusion that the seed of the disease is primarily in the blood-vessels and the blood, with secondary degenerative changes in the parenchyma, and that the cause is either some microbe or toxic substance, or both."

If such a conclusion be as nearly unanimous among the leading neurologists and pathologists as this author would assume, the following cases will, perhaps, prove food for their further reflection and, possibly, a basis for the development of some new hypothetical theory relative to the satisfactory results that have been obtained by eye-treatment alone.

It is certainly inconsistent with any such visionary theory (although based upon the results of pathological and bacteriological investigation) that many cases of the most aggravated and chronic type of chorea should get well without recourse to drugs and remain free from any return of spasmodic movements of the face, trunk, or limbs, whenever sources of peripheral reflex irritation are scientifically investigated and satisfactorily corrected.

If it can be shown (as it has already been to the satisfaction of many) that a thorough relief from eye-strain has alone produced a complete and permanent cure of many cases of chorea that were chronic and, apparently, hopeless, I doubt if the

---

[1] American Journal of the Medical Sciences, January, 1894.
[2] Italics my own.

(106)

majority of intelligent practitioners of medicine will concur in the belief that a purely local treatment (confined to the muscles of the orbit and the refraction of the eyes) effected such results through elimination of microbes or by neutralizing toxic materials.

It was not long ago that a few enthusiastic pathologists became strongly imbued with the belief that some intestinal poison was to be discovered as the cause of epilepsy,—a theory that, to my view, has not to-day the support of the best minds in the profession. It utterly fails to explain the published results obtained by me in twenty-six aggravated cases of that disease, treated exclusively through the eyes,[1] as well as similar cases that have been reported by others.

Neither has the theory that a rheumatic diathesis creates chorea, which was for a long time highly extolled, any practical bearing upon the relief of choreic symptoms, in the vast majority of the cases of chronic chorea that are encountered. At the present day, as in decades past, the physician continues to treat choreic symptoms with arsenic, as about the only drug that seems to control them. The changes are rung, from time to time, in very obstinate cases, by incorporating with arsenic, various tonics, strychnia, cold packs, codliver-oil, chloral, morphia, picrotoxin, cimicifuga, antifebrin, salol, various electric currents, experiments with magnets, and such other steps as the medical attendant (in his dire extremity) deems best suited to the exigencies of the case.

Some choreic patients seem, for a time, to recover under each of the various forms of medication suggested.[2] Some get well only when all of the drugs mentioned have been withdrawn; but a few continue to exhibit their spasmodic contortions in aggravated forms until, wearied with attempts to obtain relief, they resign themselves to fate or, in their despair, are led to search intelligently for sources of reflex irritation.

We are all exposed (from birth until death) to the presence of certain atmospheric germs, or microbes. Many forms of these are constantly being introduced into our bodies through the air, our food, and the water we drink. Constantly, through

---

[1] See Chapter VI. In these pages I have given in full the various steps taken during the treatment of each case. The reader is referred to this contribution for many points that bear upon the diagnosis and correction of heterophoria.

[2] There is scarcely a drug in the materia medica that has not been, at some time, employed with apparent benefit in the treatment of chorea.

life, are certain toxic materials also being formed within the
human body and eliminated through the kidneys. liver, intestinal tract, etc.

In some subjects, when the most favorable conditions for
their retention. absorption, and multiplication exist. do these
microbes and toxic substances hasten the dissolution of life;
while in others. that possess a reactive power capable of combating the elements of disease, life goes on, in spite of microbes,
with its normal growth and decay.

We are forced to admit that seeds sown by the farmer, as
well as those of the weed carried and distributed by the wind,
flourish best in soil that is properly prepared for their reception
and development.   Is it, therefore, irrational to assert that any
of the underlying causes of nervous debility. any constant
sources of excessive and unnatural nervous expenditure. any
hereditary condition that creates an uninterrupted leakage of
nerve-force, may be an important factor in fitting the human
soil for the seeds of disease that are ever present and ready to
take root?[1]

Should investigation end whenever microbes or toxic materials are discovered in the human economy? Should the causes
of a diseased condition be regarded as determined so long as
the underlying factors that enable this cause to exist and flourish
are ignored? ·Which is the most important in medicine,—to
ascertain what adventitious agent can do damage. under favorable conditions, or to so improve the nervous power of a patient
as to enable the body to resist the seeds of disease?

So long as we are incapable of fighting our foe (when once
established with a firm foothold within the human economy).
what have we gained, from the stand-point of the sufferer. by
simply knowing the name, exact appearance. or seat of the
enemy?

I would not be construed as decrying the scientific results
obtained by careful. painstaking bacteriological or pathological
research. They undoubtedly tend to furnish the medical practitioner with some knowledge, after the chaff has been eliminated and the wheat carefully collected, that may. in time. aid in
the prevention of disease and, possibly, facilitate its cure.   But,
while these investigations are going on, while with great rapidity

theories are being hatched and exploded, while the views of to-day may not be those of to-morrow, is it not the duty of every earnest practitioner of medicine to listen with attention to clinical facts that are capable of absolute verification, that are often fraught with immediate results to suffering humanity, and that open a wide field for the treatment and relief of obscure nervous diseases?

With these few preliminary remarks I approach the discussion of certain factors in the causation of chorea that I believe to be vitally important. I shall endeavor to demonstrate the accuracy of these views by the details of treatment of several illustrative cases of extreme and chronic chorea.

In the first place, I would assert (as I have often done prior to this date) that clinical experience has demonstrated most positively a *direct causal relationship between eye-strain and chorea.*

The forms of defect encountered in the orbit, and their degree, also, are (in a very large proportion of chronic and transient cases of chorea) such as to entail upon the sufferer an enormous expenditure of nerve-force that is not only unnatural and detrimental to health, but is capable of arrest by skillful hands.

This assertion is not a new one. It has been made over and over again, during the past decade, by many oculists of repute. It is a statement that can easily be verified by any one who practices the later methods of investigating refraction or anomalies of adjustment of the muscles of the orbit (heterophoria); yet it is one that is to-day too commonly ignored by many neurologists. Some seem to be too deeply engrossed in hunting for obscure causes of disease to see those that are already demonstrated, while others, apparently, prefer to accept any visionary theory rather than a plain, clinical fact, if that fact tend to relegate drugs to the background.

To any one who delves in the medical literature of the past, and who reads the theories that have flourished and been forgotten, it becomes at once apparent that the tendency of this age is to seek for the causes of disease, and, as far as possible, to remove them.

The argument that is advanced here is no longer a theory. It is supported by well-established physical laws and sustained by physiology. Furthermore, the results obtained when the abnormal eye-factors have been scientifically determined and removed can be shown to be almost instantaneous, and somewhat startling, in many of the cases thus treated.

Some years ago one of the leading medical journals of this country published in its editorial pages,[1] in strong advocacy of this method of treatment, several columns under the following heading: "A Great Medical Discovery Ignored." From this editorial I quote the following paragraphs:—

"There are few medical truths that have been discovered fraught with more possible and incalculable good to humanity than one that is ignored by the great body of the medical profession.

"There are explanations and sufficient reasons for this anomalous fact. Among them may be noted these:—

"1. The discovery has come about slowly and silently. It has been made by no one man, and has come with no flourish of international congressional trumpeters. So softly and slowly has it crept into scientific medicine that its own advocates are but half-aware of it, and do not yet realize its almost unparalleled value.

"2. It is a therapeutic measure that depends for its exercise upon an exactness of knowledge of delicate, mysterious physiological and psychological functions that few possess, and upon a subtle discrimination and judgment with which, by character or education, few are endowed.

"3. It has the misfortune to depend for its promulgation and practical application upon the specialist, and almost upon the specialist of a specialty,—and this in a profession and in an epoch in which it is fashionable to sneer at specialism and at the specialist who dares plead for the truth he knows,—and that, at first at least, only he can know."

Perhaps the most common experience that I personally encounter in my office is to have patients tell me that either their doctor or some oculist whom they have consulted has said to them that "their eye-trouble is the *result* of their physical weakness and *not a cause ; that the relief of the eye-trouble can have nothing to do with their recovery ;* and that all statements to the contrary are not supported by facts."

It is for the purpose of demonstrating the counter-propo-

---

sition. of showing that eye-strain may indirectly be the *cause* of obscure disease and not its result. and of turning. if possible. the channel of medical thought so as to benefit suffering humanity, that I deem it wise to raise my voice here in defence of a method of treatment that will often accomplish what drugs will not, and that is based upon science rather than therapeutic speculation and empiricism.

The cases of chorea which I report here in detail I have selected from a number on my private records. in order to demonstrate most positively my view of the points at issue. They have been seen by many physicians from time to time. To many of these patients the verbal or written opinion of prominent medical men had been given prior to my seeing them, "that *organic disease* unquestionably existed, and that the eyes had nothing to do with the causation of the symptoms."

In some of these cases no possible explanation of the facts here recorded can be suggested, except to admit that the correction of an existing eye-strain caused a cessation of the leakage of nervous force that had been going on for years; and, by so doing. the sufferers had been enabled to regain their normal condition.

The sad results. viewed from the stand-point of suffering humanity, that are entailed by indifference and prejudice in men of scientific reputation cannot be estimated.

By giving expression to others of their opinion concerning what they have not properly and patiently investigated themselves or will not see, thousands of sufferers are doomed to a life of misery.

Such patients naturally believe that abnormal eye-factors in their own case have been sought for by the latest methods and found to be absent by one who stands high in his profession. They quote to their friends the positive assertions of him (whom they believe infallible). that " eye-muscles are not worth investigating "; that " all deviating tendencies of the eyes are invariably due to errors of refraction "; that " Javal's instrument has been used, and that settles the question forever "; and other similar expressions, indicative either of inexperience, bigotry, or prejudice, that too often come to my own ears.[1]

[1] The general reader should consult those pages in Chapter II that relate to these various tests, in order to appreciate the clinical and diagnostic importance of each.

We are now prepared to pass to the consideration of individual cases of chorea. These histories demonstrate beyond cavil, I think, that a direct relationship between eye-strain and extreme forms of nervous disturbances can exist; that the correction of eye-strain may be followed by very marked benefit in some instances; that it is the duty of a physician to have the eyes of all patients afflicted with abnormal nervous disturbances examined early by some oculist who is familiar with and employs the latest methods; and that errors of adjustment of the eye-muscles are quite as important to detect and rectify as are marked errors of refraction.

Respecting the *relationship of chorea to anomalies of the visual apparatus*, I am led by my experience to draw the following conclusions:—

1. Choreic subjects belong to one of two classes: (*a*) those who tend to get well under almost any treatment or even without treatment and (*b*) those who fail to get relief from any medicinal aid. The latter tend to run a chronic course, usually one of unfavorable progression.

2. The *chronic form of chorea is one of the most serious and hopeless of nervous maladies,* when treated by drugs alone. It is not infrequently associated with epilepsy or with mental impairment later on in life. Chronic sick headache may often be developed, and sometimes asthenopic symptoms are very prominent.

3. Both forms of chorea are based, as a rule, upon a well-marked *neuropathic* or *tubercular predisposition.*

4. The pathology of chorea is not known. No one has ever proved that it was a "constitutional disease" in the sense that an organic lesion was essential to its development.

5. The *percentage of hypermetropia (usually latent) in choreic subjects is extremely large (apparently about 70 per cent.).* I make it a rule, with few exceptions, to correct the total error of refraction in choreic subjects fully by glasses. Often atropine has to be instilled into the eyes at frequent intervals, for a time, in order to arrest ciliary spasm and enable the patient to accept a full refractive correction without marked discomfort.

6. The glasses ordered for choreic patients should be *most carefully fitted to the face* and *accurately centered to the pupils.*

7. An *investigation* for *latent heterophoria*[1] should always be made in choreic subjects with the greatest of care and patience.

8. The *relief* of *marked heterophoria* should be finally attained only by *graduated tenotomies* upon the muscles exhibiting abnormal tension, or by an *advancement of the tendons of the muscles exhibiting defective power.*

9. *Prismatic glasses are not curative.* They should not, as a rule, be prescribed for constant use.

10. Choreic subjects are usually rapidly cured by eye-treatment alone. So large has been the percentage of recoveries to the total number seen by me, during the past ten years, that I have come to regard the prognosis of chronic chorea as extremely favorable.

11. The eye-problems encountered in choreic subjects are not, as a rule, as complicated and difficult to solve as those of epileptics; nor, in my experience, is the duration of eye-treatment apt to extend over as long a period before very decided benefit is observed.

12. I believe that the spasmodic movements which accompany and indicate organic lesions of the brain—as, for example, those of a leptomeningitis—exist in but a small proportion of choreic subjects, and are usually associated with other evidences of disease.

That organic lesions of the brain and spinal cord may produce choreic movements of an extreme type cannot be denied; but it is wise, in my experience, to be slow in giving an unfavorable prognosis in any case until all possible reflex causes of chorea have been thoroughly investigated and corrected.

13. The removal of young choreic subjects from school, or of adult choreic patients from business, is a step commonly taken by most physicians while treating chorea by drugs.

It must be apparent to all thinking minds that the rest thus given to the eyes and nerve-centers is a factor in the recovery of acute cases of chorea that is as important clinically as the drugs employed.

---

[1] A term that covers all forms of maladjustment of the muscles of the orbit. See table on page 35.

CASE 1. *Aggravated Chorea, with Impairment of Speech and Walking.*—Miss L., aged 15. Referred to me on July 5, 1893, by Dr. Symmes, of Cranberry, N. J.

Family History : The mother has always had violent sick headaches. One maternal aunt has severe sick headaches. No phthisis in the ancestry.

History of the Case : The patient has had marked chorea, off and on, for the past eight years. The first attack followed inflammatory rheumatism, but she has had no evidence of rheumatism since that attack.

About two months before I first saw her she began to have pain in her eyes and sensitiveness to light. This was followed by the worst attack of chorea that she has ever had.

The twitching of legs, arms, and face was very marked, and her speech was so thick as to be almost unintelligible. She walked with great difficulty, being unable to control her limbs.

During her first visit to the office she tore a handkerchief into threads. She twitched so violently that her head had to be held by main force during the eye-tests, while her arms and legs were moving in all directions. She had slept very poorly for several weeks.

Eye-defects : At the first examination the patient showed : Adduction, 33° ; abduction, 7° ; no hyperphoria ; esophoria, 15° ; in accommodation. esophoria, 10° ; homonymous diplopia with a red glass ; under homatropine, hypermetropia ($+ 0.50^s$) and astigmatism ($+ 0.50^c$; axis, 90°) in each eye.[1]

After wearing a 5° esophorial prism for a short time she showed esophoria of 22° and abduction of 2°.

Treatment and Results : A full correction was given for her refractive error both for reading and distance. Graduated tenotomies were performed on the internal recti,—one on the right internus and two on the left internus. Her tests two weeks after the last operation were : Adduction, 47° ; abduction, 8° ; no hyperphoria ; esophoria, 1° ; vision, $\frac{20}{15}$ in each eye, with refractive correction.

Two days after the first operation the patient's mother reported that she had slept through two entire nights very

---

[1] The high abduction of this case in connection with such extreme manifest esophoria and homonymous diplopia, when a red glass was employed, is a somewhat interesting feature in these tests.

quietly. After the second operation she appeared much more quiet and continued to sleep well.

Two weeks after the last operation and five weeks after the first visit her mother reported that the patient was "absolutely well of all choreic symptoms."

A letter from her pastor, dated October 6, 1893, says: "You ask what I think of Miss L.'s improvement. It is little short of miraculous. Both she and her parents owe you a debt that no money can discharge. When I see her now, and consider her condition three months ago, my heart goes out in thankfulness to God that he has given unto men wisdom and power to so relieve the sufferings of mankind."

Up to the writing of this chapter, a report states that no return of the chorea has been observed, and that the patient continues to exhibit perfect health.

The complete and rapid recovery of this girl was naturally a source of great surprise to every one who knew her. Her defective articulation and extreme chorea justified alarm, and a strong suspicion that some organic disease had been developed, of which the chorea was but one manifestation.

CASE II. *Chronic Chorea, Followed by Loss of Power in Both Legs and the Right Arm.*—Miss S., aged 10,—a large child for her age,—was sent to me for treatment on June 15, 1891.

This case is one of the most remarkable ones that I have ever seen. There was every reason to think that an organic lesion of the brain existed. I could hardly believe, at first, that marked benefit from eye-treatment could result.

Family History: Mother is perfectly well. Father has had some eye-trouble. One cousin on maternal side had epilepsy. One maternal aunt has nervous prostration. Paternal grandmother had nervous prostration. Considerable phthisis in the father's family.

History of the Case: The patient has had several severe sicknesses from acute diseases incident to childhood.

Eight months ago it was noticed that she could not keep her hands still, and was constantly knocking things over. This kept growing worse until, six weeks before she came to me, she lost all use of the right hand. Her ankles, which had been growing weak for some time, turned over so badly that she occasionally fell down. At times her speech was badly affected, and her words often ran together so as to be almost unintelligible. When she reads the letters blur badly and her eyes become very much inflamed. Her eyes had troubled her for some time before any choreic symptoms were noticed. One eminent neurologist of this city had seen the patient and pronounced the case one of organic brain disease. He had given a very unfavorable prognosis.

Eye-defects: Latent hypermetropia, $+ 1.00^s$; esophoria, 8°; adduction, 38°; abduction, 2°; right sursumduction, 2°; left sursumduction, 2°. A high degree of latent esophoria was subsequently disclosed by this patient.

Treatment and Results: The treatment consisted in giving prisms, bases out, over each eye for constant wear, and later in performing graduated tenotomies upon both internal recti.

The improvement in the choreic symptoms began almost as soon as the prisms were put on. At the end of the first week her mother reported that she could dress herself much better than formerly; that her right hand rested quietly in her own when walking, whereas it used to be impossible to hold it, on account of the twitching. She also began to use her right hand for the first time while in my consultation-room, to denote the position of test-objects while her eye-muscles were being tested.

The patient, four months after treatment was begun, wrote and sewed with her right hand, walked perfectly well without turning of the ankles, and had no choreic symptoms whatever. The power in her right arm was fully regained later. The parents state that " she is as well as she ever was," and express the greatest gratitude because of her restoration to health. For nearly six years no return of her choreic symptoms has been observed.

On January 19, 1897, a friend of the parents of this patient visited the office. He reported that " the patient is now a strong and robust young lady, and that she had never, at any time, had the slightest symptom of her old trouble."

Since then I have seen one of the parents and a confirmation of the complete recovery of this patient was obtained. Every choreic symptom disappeared over five years ago, and no relapse has occurred since that time.

CASE III. *Aggravated Type of Chronic Chorea, Accompanied by Deformity, Headache, Asthenopia, and Inability to Work.*—Miss C., aged 26. Referred to me for diagnosis on February 27, 1888.[1]

Family history not known.

History of the Case: The patient was a poor factory-girl that was sent to me by a medical friend, Dr. O'Connor, of Holyoke, Mass., to see if anything could be done for her. No one who had seen her could make a diagnosis, although she had been seen by most of the prominent physicians of the State. Since her tenth year she had suffered with neuralgic headaches, asthenopia, and persistent trembling in her hands and arms. For four years there had been a stiffness and rigidity of the neck, accompanied by severe pain in the neck and "choking-spells" when she attempts to speak or when excited. There is also marked contracture of the hands and arms.

Her appearance when she came to me was one of remarkable deformity. Her chin seemed to be held firmly in the region of the chest, at the level of the fourth button of her waist. Both upper limbs were contractured, in the state of semi-flexion, at elbows, wrists, and fingers, and trembled excessively when she tried to use them. They could not be extended, nor could the patient make use of the hands. She had been obliged to give up work on that account. Whenever she talked her face became painfully drawn and distorted. The mouth would especially be drawn downward. Under the least excitement she would be seized with what she called "choking-spells." The throat-muscles would contract and interfere seriously with respiration. She had constant headache, chiefly in the forehead and occiput, and a persistent pain in the neck. No painful points existed, nor did pressure, movement, or other tests reveal the existence of organic disease of the spinal cord. Sensibility to touch, pain, and temperature was normal. Motility was unimpaired. The muscular sense was perfect. No inco-ordination existed. I considered the case one of aggravated chorea, complicated by headache and asthenopia.

Eye-defects: Hypermetropia, + 0.75; esophoria (mostly

---

[1] This case has been previously mentioned in Chapter III because headache and intense neuralgic paroxysms constituted an important part of her clinical history. The case is one of sufficient interest, however, to warrant repetition here, as the patient was exhibited at medical societies, before I saw her, as a medical curiosity.

latent), 20°; left hyperphoria, 3°; adduction, 22°; abduction, 5°; right sursumduction, 2°; left sursumduction, 5°.

Treatment and Results: The muscular anomalies were satisfactorily relieved by graduated tenotomies of both internal recti and the left superior rectus. Applications of static sparks were made daily for about two months. The patient had, by this time, regained the use of both hands, carried her head erect, had no headaches and little spinal pain. For awhile she returned to work. The effect of hard labor brought back her "choking-spells"; hence she was advised to take absolute rest if they continued. She now has some asthenopia remaining and an occasional headache. Her facial contortions persist to a slight extent when she is unduly excited; but otherwise no spasmodic movements are noticed. I suspect that some latent hyperphoria exists still, and that further operative work on the eye-muscles will be demanded before the patient recovers completely.

The last report from this patient stated that there was a steady improvement in her condition and that she had again returned to work. She had no headache and few "choking-spells." Her head and hands were no longer affected with distortion.

CASE IV. *Chronic Chorea of Twelve Years' Duration, with Horrible Facial Contractions.*—Mr. C., minister of the gospel. Mother died of phthisis.

This remarkable case is quoted from my work on nervous diseases.[1] This patient was seen by me in the office of my friend Dr. George T. Stevens.

Family History: No nervous diseases had existed among his direct ancestors or in remote branches of his family.

History of the Case: About twelve years before this interview his family had noticed frequent facial contortions which he was unable to control. A trip to Europe, and parish labors in a district where he spent most of his time in a carriage, and wrote but little in his study, prevented its increase for about four years. He then became the pastor of a church and began active labor in his study. The facial contortions grew rapidly more aggravated in character. Every feature would become horribly distorted; the eyes would close, the forehead become

---

[1] Lectures on Nervous Diseases, The F. A. Davis Co., Philadelphia, 1888.

terribly wrinkled, and the nose and mouth would assume attitudes which no one could possibly imitate by volition, and which are difficult to describe. The hour of retiring was particularly dreaded, because the facial spasms would become terribly persistent and severe as soon as the eyes were closed and a recumbent posture was assumed. The facial contortions were always least severe in the morning, and grew more severe as the day progressed. No medicinal treatment had ever benefited the patient.

Eye-defects: An examination showed hypermetropia of a high degree, esophoria of 6°, and hyperphoria of right eye of 3°. He had been wearing *prisms with the base inward*, at the suggestion of an oculist. Naturally he had been made decidedly worse by their use.

Treatment and Results: A partial tenotomy was first performed to correct the hyperphoria. The facial spasms *ceased within an hour*, and *no sign of chorea was observed for two entire days*. On the third day a very slight twitching about the mouth developed. A partial tenotomy of the internal rectus of the left eye was then performed. This completely corrected the esophoria.

Subsequent to the second operation the patient had few, if any, choreic movements. He stated to me that, "unless excited, his face remained absolutely quiet." and that, for the first time, he had that day "been able to attend a meeting of ministers and look them in the face without facial spasms while discussing church matters." During his recital of his various symptoms, etc., to me. his face only showed one very slight convulsive movement. Thus, in less than one week, were the convulsive spasms of his face almost completely arrested by correcting a hypermetropia and two muscular defects associated with the eye.

Such a case is rarely encountered. The patient was an adult. The duration had exceeded twelve years. The spasmodic movements were terribly severe. All medication had failed even to ameliorate them. They became greatly aggravated as soon as the patient was compelled to use his eyes in study or writing. He could not even "look out of a car-window" without being thrown into a most distressing state; yet, in spite of all these unfavorable facts he was apparently perfectly well when I last conversed with him.

CASE V. *Extreme Chorea of Face, Arms, and Legs.*—
Miss H., aged 15, referred to me on August 17, 1889, by Dr.
G. B. Banks, of Huntington, L. I.

Family History : The mother has very severe headaches.
One brother has sick headaches. One sister had phthisis.

History of the Case : The patient had been perfectly well
until she had pneumonia, eighteen months prior to her visit to
my office. About six months prior to the first examination by
me she began to have twitchings in her hands and arms. These
grew gradually worse until, at the present time, they affect her
legs, arms, and face to a marked degree. She frequently drops
whatever happens to be in her hands, her mouth twitches very
badly, and she is extremely nervous and excitable. She had
some asthenopic symptoms four years ago, but has not been
troubled since with her eyes. She is also troubled with marked
dyspnœa on the slightest exertion.

Eye-defects : Under atropine the patient showed : Hypermetropia, $+ 0.75^s$; adduction, $20°$; abduction, $6°$; sursumduction, right, $2° +$; left, $2° +$; no hyperphoria ; esophoria,
$8°$. Two days later, while wearing a $2°$ esophorial prism, she
disclosed $10°$ of esophoria, with an abduction of $2°$.

Treatment and Results : The patient was at once given $4°$
of prism, base out, for her esophoria, and she wore them constantly for about six weeks, with some slight improvement in
her symptoms.

September 4, 1889. A graduated tenotomy was performed
on the right internal rectus. During the next two weeks the
choreic movements had almost entirely ceased.

October 8th. A graduated tenotomy was performed on
the left internal rectus. Two weeks later her father reported
that she was going to school without bad effects, and was rapidly gaining flesh and strength. One month later he said
that her dyspnœa had entirely disappeared.

November 11th. Patient reports that, prior to treatment,
she used to throw her pencils on the floor without knowing
why, but has not done so since. She has no headache after
study, as formerly, and has gained fifteen pounds in weight.
The teachers in school have spoken to her parents of her
marked improvement.

December 21st. The right internal rectus was again divided, as some latent esophoria had been disclosed.

October 9, 1890. Patient reports that she has had no chorea and only an occasional headache.

Her muscular tests are: Adduction, 20°; abduction, 7° +; sursumduction, right, 2° +; left, 2° —; right hyperphoria, ¼°; esophoria, 1°.

Reports from time to time, during the past three years, have been received from a neighbor, although the patient has not visited my office during this interval. From what I can learn, this patient continues to be absolutely free from any evidences of chorea.

CASE VI. *Terrible Attacks of Neuralgia of the Stomach and Persistent Trembling of the Head, Face, and Limbs. Alarming Attacks of Suffocation.*—Mrs. G., aged 42. Referred to my care on November 10, 1886. This case was one of the most surprising, both to myself and the patient, that I have ever encountered. The long persistence of the symptoms, the extreme physical weakness, the alarming attacks of laryngeal spasm that would occur, without warning, at intervals, and the tremor of the entire muscular system that persisted whenever any excitement or fatigue occurred, tended to make the prognosis grave and to justify a strong suspicion of organic disease.

Family History: Several blood-relatives died of phthisis; father and brother died of phthisis.

History of the Case: This patient had been for sixteen years a chronic invalid. She was unable to bear the least excitement. Even the companionship of her family for an evening was at times too great a strain upon her nervous system. She was at times a great sufferer from severe paroxysms of neuralgia of the stomach, and frequent attacks of alarming shortness of breath and a sense of impending suffocation would occur. I personally witnessed one of these attacks in my office. and it was entirely free from a trace even of hysteria. It was of much shorter duration than an asthmatic attack, and seemed to be due to a spasm of the larynx. She became markedly cyanotic and suffered alarming shortness of breath.

In addition to these symptoms, this patient suffered for many years from an *uncontrollable trembling of the facial muscles and limbs, when at all startled or excited.* She had been for years unable to attend places of amusement or to bear physical exertion.

Eye-defects: The patient was found to be emmetropic (when under atropine). Esophoria (manifest) of 3° existed ; abduction, 5°. A high degree of latent esophoria was disclosed later, with a proportionate decrease in the abduction. Homonymous diplopia at times existed prior to operation.

Treatment and Results: Much to my surprise (as she had a marked phthisical history), an examination of her eyes showed no refractive error (even when under the influence of atropine). She showed, however, a very high degree of esophoria, and a graduated tenotomy was performed upon both of her interni. The effect was magical. She recovered her health completely within two months, and is to-day able to endure as much as when a young girl. One of the last reports from her, some time ago, states that she had "shopped all day and attended the theater in the evening." An old friend of the family lately alluded to the case, in my presence, as one "not of cure, but of resurrection."

Ten years have now elapsed since the patient was relieved of her eye-strain. During this period no return of her old symptoms has occurred, nor has she been obliged to resort to drugs for the relief of any physical ailment. At the last tests this patient showed: Adduction. 33° ; abduction. 8° ; no esophoria ; no hyperphoria.

This case is very interesting (from the stand-point of the oculist) in that there was absolutely perfect refraction, even under atropine, and that the esophoria was almost totally latent. The remarkable results obtained by eye-treatment have remained undiminished during a period of over seven years, in spite of extreme physical exercise, as compared with her powers of endurance prior to the eye-treatment.

CASE VII. *Choreic Movements of the Head, Face, and Limbs, with Recurring Attacks of Gastric Pain, Pallor, and Extreme Weakness.*—Master R., aged 10 years. Referred to me on November 10. 1888.

Family History: Father for thirty years suffered from most extreme attacks of sick headaches every week, until his latent hypermetropia was corrected. Phthisis is extremely common among the paternal ancestry; one maternal uncle was insane. Almost all the paternal ancestry has suffered from sick headache.

History of the Case: This patient was the son of a physician, who had himself been cured of terrible sick headaches and complete nervous prostration by the correction of a latent error of refraction that had not previously been suspected by himself. The son had been somewhat delicate from youth, being highly nervous and generally anæmic. At frequent intervals severe gastric pain would appear, without apparent cause, and create symptoms of extreme pallor and faintness.

When about 6 years of age mild symptoms of chorea appeared. The child would twitch and jerk his shoulders, face, and limbs, when at all excited. As he grew older these choreic symptoms increased to such an extent as to cause an alarm and to justify an examination of the eyes as a possible factor in causing the chorea. At the time when I first examined the boy he could not attend any place of amusement without attracting attention by his choreic movements, and at times they were very marked, irrespective of excitement. Medication had accomplished nothing. The patient had been frequently removed from school, because it was found that study aggravated the chorea.

Eye-defects: Under atropine this patient showed: Latent hypermetropia of 2 diopters; esophoria (manifest) of 4°; abduction, 4°. Later a high degree of latent esophoria disclosed itself, under the influence of prisms, with homonymous diplopia and no abduction.

Treatment and Results: The child was ordered $+1.00^s$ glasses for constant wear. Three graduated tenotomies were performed upon the interni at intervals of about two months, as latent esophoria of a high degree was disclosed, from time to time, by the patient.

The result was very gratifying. The child recovered completely from his chorea within a year and has never shown any return. Within the past five years he has developed into a very tall and strong boy, weighs 150 pounds, and is in perfect health. He uses his spherical glasses now only for study, reading, and any close work for the eyes. He has attended school regularly for the past seven years without any ill-effects upon his health.

CASE VIII. *Aggravated Type of Chorea Affecting the Face, Arms, and Legs. Marked Impairment of Speech.*—Miss B., aged 18. Referred to me on December 1, 1892.

Family History: Mother is very nervous and excitable in temperament.

History of the Case: This was a well-developed young girl. Up to within a year of her first visit she had been considered to have a nervous temperament, but had shown no symptoms of chorea. At this time she was allowed by her mother to visit a school-friend for a few days. During this visit she participated in various amusements and lived a life different from the routine life which she led at home. On her return from this visit her mother noticed that she was unnaturally excited and extremely nervous. She twitched her arms and face in a very peculiar and unnatural manner, would cry and laugh hysterically, and did not articulate as distinctly as before the visit. These symptoms seemed rather to increase than decrease within the next few days, and caused great alarm to the family. Various forms of medication were tried, without benefit, during the next few months. The choreic symptoms steadily increased in intensity and marked physical weakness appeared. She became unable to play upon the piano, could not bear the excitement even of girl companions, and her speech became so thick and indistinct as to become almost unintelligible, at times, except to those who knew her well. A walk of one or two blocks would tire her so that she would have to lie down and rest.

At her first visit to my office her choreic movements were very severe in her arms, legs, and face. It was impossible for me, at times, to understand what she said, as she sputtered and talked as though she had a foreign body in her mouth. No evidence of organic disease could be found, and there was no apparent abnormality in the perception or localization of sensations of touch, pain, or temperature. The power of individual muscles seemed to be intact. The mother had become strongly impressed with the idea that her daughter was suffering from some organic trouble and that the prospect of recovery, under any treatment, was extremely discouraging.

Eye-defects: Under homatropine she showed hypermetropia, $+ 0.75^{x}$. Her muscular tests were: Adduction, 15°; abduction, 9°; sursumduction, right, 3°; left, 3°; no hyperphoria; esophoria, 1°. Later on she disclosed a high degree of latent esophoria.[1]

---

[1] This case illustrates a latent esophoria of a high degree, with an apparently *high abduction* at the first interview.

Treatment and Results: This patient showed so little eye-defect at the first visit that the case did not seem a favorable one for eye-treatment. Prisms were given, however, for esophoria, and increased in strength as fast as she showed any surplus that would justify such an increase. In one week, while wearing 6° of prism, she showed esophoria of 7° and abduction of 3°. A graduated tenotomy was then performed on the right internal rectus.

Her mother reported, the next day, that she was better than at any time for a year. Two weeks later, as 7° of latent esophoria had become manifest, a graduated tenotomy was performed on the left internus.

The improvement in the choreic symptoms was noticeable within a week after this operation, and she grew steadily better, so that, at the end of a month, she was apparently well. No choreic movements were noticed, her speech was distinct, she walked several miles daily, could dance, and play the piano as formerly. When last seen, on December 1, 1893, one year from the date of the first visit, there was no return of her old symptoms. Her eye-tests at that time were: Adduction, 25°; abduction, 9°; no hyperphoria; no esophoria.

Case IX. *Chronic Chorea of Thirty-one Years' Duration, Affecting the Head, Face, and all the Extremities.*—Miss D., aged 33. Referred to me for diagnosis and treatment on May 27, 1887. This case is one that I reported some years ago, in my work upon nervous diseases,[1] with two illustrative photographs that showed the changes in expression and the decrease in choreic movements caused by the relief of heterophoria by graduated tenotomies. I quote this case here because the chorea was very extreme and had persisted for over thirty years, in spite of all therapeutic treatment and the best medical advice.

Family History: The father had pulmonary hæmorrhages for many years. One paternal aunt died of "hasty consumption." Sick headaches are very common among both paternal and maternal ancestors. Neuralgia is a frequent complaint among the paternal ancestors.

When 2 years of age this girl developed chorea. The spasmodic twitchings steadily grew worse, in spite of the fact that her father was a physician and that she had the services

---

[1] Lectures on Nervous Diseases. The F. A. Davis Co., Philadelphia. 1888.

of the most skillful medical men, from time to time. The twitchings began on the right side, but they subsequently involved the left side and also the head and face.

She had suffered some from sick headaches, as had also her sister. The *hands have gradually become so contractured* that all attempts to use them are more or less distressing. Her fingers could not be extended farther than would suffice to grasp small objects.

When I first saw this patient she was unable to write except by grasping the pencil with all the fingers and the palm of the left hand and holding the left hand with the right hand as the spasmodic movements of writing were made. She walked with a peculiar, unsteady, and crab-like gait, ate with difficulty, and suffered great pain between the shoulder-blades and over the first lumbar vertebra (two points, by the way, which are very frequently attacked, in my experience, when eye-strain is present). She had never written with ink. Prior to menstruation (which occurred at 17 years of age) the patient had experienced attacks (probably epileptic) which she describes as "those of numbness, followed by a loss of consciousness." She has had chronic constipation all her life. The memory and mental faculties are perfect.

When I first saw this patient the spasms were very violent, especially about the face and neck. The limbs were jerked about, the fingers too tightly clenched to grasp anything, and the speech was rendered peculiarly spasmodic and almost unintelligible, at times. She sputtered and frequently ejected drops of saliva when endeavoring to converse.

Eye-defects: At the first examination she exhibited no refractive error; but under atropine a moderate degree of hypermetropia (1.75 D.) was detected and spherical glasses (+ 1.00 D.) were at once provided. In order to test her eye-muscles, the services of Prof. J. Williston Wright, of New York City, who saw her with me by invitation, were invoked to hold her head. This he did, with no small effort, by clasping the head on either side and firmly pressing her head against his body as he stood behind her chair. During this examination she whistled shrill notes on two occasions and underwent the most violent facial and bodily contortions.

Treatment and Results: The results of this imperfect examination (necessarily so, under such conditions) indicated to me that a high degree of hyperphoria existed, and, as I could not again see the patient for some months, I decided to perform a free, but incomplete, division of the left inferior rectus muscle. I then instructed the patient to try and get a photograph taken, if possible, before she saw me again. She laughingly said that she had never been able to have a picture taken, but she would do so if she could. She then departed for home with instructions to return to me for treatment in the autumn. The first picture received of this case was one that she was able to have taken three weeks after the operation, when her head and shoulders had become comparatively calm, as a result of the relief afforded by it. This photograph was deemed, at that time, a great success by herself and friends. In it the blurred outlines, which indicate that the movements were still somewhat active, are very apparent.

During the next autumn this patient was under my care for some eight weeks. I partially divided the right superior rectus and both externi, to overcome a high degree of right hyperphoria and exophoria, and administered static sparks daily to the spine and limbs. A second picture that I possess gives, better than words can describe it, an idea of the wonderful improvement which has taken place. Prior to her departure for home she could thread the finest cambric-needle and pass her fare to the conductor of a street-car without attracting the notice of passengers or throwing it out of the window, as she certainly would have been apt to do two months previously. She could fully extend her fingers, walk several miles a day, write with far greater certainty and ease, and eat at a boarding-house table without exciting comment. Her limbs still twitch somewhat immediately before going to sleep, and, in the presence of strangers or when unduly excited, she still shows some spasmodic movements of the face and shoulders. When calm she is, however, perfectly composed and almost entirely free from convulsive movements. She considers herself as practically cured; but I suspect that time and some further operative work upon the eye-muscles will be demanded before complete restoration to health is effected.

During the past seven years this patient has not been able to come to New York to complete the work that I began upon her eye-muscles. She has, however, improved steadily in her physical and nervous condition. She writes me several letters each year, in ink, and reports that she continues to be practically free from all chorea, although still somewhat nervous when tired or excited.

As I regard this case as one of the most distressing and typical cases of chronic chorea ever reported, it may be well to state that the patient is well known to Prof. A. M. Phelps, M.D., and Prof. J. W. Wright, M.D., of this city, and Professor Woodward, M.D., of Burlington, Vt. She has also been seen by many members of the profession, both from this city and distant States, during her treatment in my office.

During the whole treatment of this patient no drugs have been employed, and the photographs that I published in my work on nervous diseases are from untouched negatives. I attribute to the static applications the rapid relief of the contractured state of the fingers and the improvement in her general strength; but, from many facts observed during my treatment of her, I am convinced that the relief of the eyestrain is alone deserving of whatever credit may be claimed for her recovery. Four weeks before she was dismissed from my care she read and sewed continuously for several days, and was immediately precipitated into a relapse, which as rapidly subsided when the cause was ascertained and its recurrence prevented.

CASE X. *Extreme Chorea, Accompanied with Complete Nervous Prostration that Entailed a Retirement from all Business Pursuits.*—Mr. L., aged 35; married; manufacturer. Referred to me on November 19, 1892, by Dr. E. W. Hedges, of Plainfield, N. J.

Family History: Father died of phthisis and mother of some kidney disease. Patient has six children, all living and well.

History of the Case : In February, 1892, the patient had a "collapse" while at his office. His face flushed, his head was confused, he was unable to attend to any business, and had to be taken home. He gave up his business, tried various tonics, change of air, etc., but was not materially improved.

At the time of his first visit to me he was absolutely unable to attend to any business; the least excitement disturbed him; he had pain in back of the head and eyes, and had marked choreic symptoms. His gait was jerky and very uncertain from loss of control of the leg-muscles. Walking tired him greatly, although he was a strong and robust-looking man. His face was in constant motion and his speech very much impaired. It was with difficulty that he was kept quiet long enough to test his muscles and eyes.

Eye-defects : Under homatropine he showed : Hypermetropia and astigmatism; O. D.. + 1.25ˣ; O. S., + 0.75ˣ ◯ + 0.50° (axis, 90°). His muscular tests were: Adduction, 20° ; abduction, 6° ; no hyperphoria ; esophoria, 2°. Later he disclosed a very high degree of latent esophoria.

Treatment and Results : All drugs were discontinued. Glasses were ordered for constant wear as follows : O. D., + 0.50ˣ; O. S., + 0.50° (axis, 90°). Prisms were given for relief of his esophoria, at first, as he disclosed esophoria to a high degree very rapidly. Graduated tenotomies were performed upon both internal recti within two weeks. During the next four months each internal rectus was again button-holed. The improvement in this patient was quite marked within a few weeks after the first two tenotomies. He was much more quiet, could walk steadily without fatigue, his speech was normal, and his choreic movements had almost entirely disappeared.

In May, 1893, he had a slight return of his old symptoms after excessive use of his eyes, and was given prisms for relief of some remaining esophoria.

When seen, in September, 1893, he was feeling well and attending regularly to his business. Since January, 1894, the patient has disclosed some latent hyperphoria. This has been corrected. The patient has improved sufficiently to resume control of his former business, and exhibits very few of his old choreic symptoms. He regards himself as practically cured. I have not examined his eyes for a year. but get very satisfactory reports from him.

CASE XI. *Chorea of Facial Muscles and Limbs, of Three Years' Duration.*—Master A., aged 11. Referred to me on October 22, 1890.

Family History : Mother had severe sick headaches until relieved by the correction of hypermetropia and esophoria under my hands. Father had chorea until 17 years of age, and has always been very nervous. He was operated upon for exophoria by me, with decided benefit.

9

Phthisis is common among both maternal and paternal ancestry. Chorea is common on the mother's side. Paternal relatives are very nervous, and one aunt has St. Vitus's dance.

History of the Case : When this patient was a baby he had catarrh of the stomach, and has had a weak stomach ever since. When 3 years old he began to have "twitchings," which gradually grew worse, and have continued, without intermission, ever since. The choreic movements are noticeable in the extremities, but especially marked in the face.

Eye-defects : Under homatropine he showed: O. D., + 0.50ˢ; O. S., + 0.50ᶜ ◠ + 0.50ᶜ (axis, 90°). His muscular tests were: Adduction, 16°; abduction, 7°; sursumduction, right, 3°; left, 3°; no hyperphoria; esophoria, 1°. After wearing prisms for several weeks he disclosed considerable latent esophoria.

Treatment and Results : A 1° prism was given for esophoria and increased in strength as fast as he showed any surplus over the prism. Two months later a graduated tenotomy was performed on the left internus. A full correction for his refraction was also given for constant wear. For some weeks after this operation there was a marked improvement in the choreic symptoms, but a relapse soon occurred, and esophorial prisms were again put on, with marked benefit.

February 26, 1891. Ten weeks after the first operation the right internus was divided.

March 5th. Patient showed no choreic symptoms and looked like a different boy. His muscular tests are: Adduction, 32°; abduction, 8°; no hyperphoria; esophoria, ¼°. Since last note this patient has been relieved of all choreic disturbances, as far as is known.

CASE XII. *Chronic Chorea with Marked Impairment of Speech and Great Difficulty in Walking.*—Miss R., aged 35; teacher.

Family History: Father was myopic and suffered with headache. Mother had St. Vitus's dance for years, and died from it. Three brothers and one sister have headaches and wear glasses. One cousin on each side is in an insane-asylum.

History of the Case : When a child this patient suffered from headache. For seven years patient worked very hard at teaching, and suffered from gastric trouble.

In May, 1892, she broke down completely with nervous

prostration and chorea. She could talk but little, and that with difficulty. Her limbs were so much affected that she could hardly control them enough to walk. She was sent to Colorado for a year, returning in July, 1893, not much improved.

At the date of her first visit to me she walked difficultly. Her speech was very slow and jerky, and her sister states that her mind is failing very noticeably. There is a fine tremor of the muscles, which is felt wherever she is touched, but no marked convulsive movement.

Eye-defects : Under atropine she showed : Hypermetropia and astigmatism ; O. D., + 0.25 $\bigcirc$ + 0.25° (axis. 90°) ; O. S., + 0.50° $\bigcirc$ + 0.50° (axis. 90°). Her muscular tests, at the first examination, were : Adduction. 16° ; abduction, 4° ; no hyperphoria ; esophoria, 2½°. Within two days, however, while wearing a 6° prism, she showed 8° of esophoria, with unconquerable homonymous diplopia. Later on she disclosed a high degree of latent esophoria.

Treatment and Results : Glasses were given for constant wear to correct her astigmatism, and within a week a graduated tenotomy was performed upon each internal rectus.

She began to improve soon after these operations ; was much quieter, talked better, slept well, and began to walk regularly every day. As more esophoria became manifest, prisms were given and gradually increased in strength. Two weeks after the first operation a graduated tenotomy was again done on the right internus.

One week later her tests were : Adduction, 26° ; abduction, 6° + ; no hyperphoria ; esophoria, 1°.

At this time she reported that she was sleeping eight hours every night, walking from two to eight miles a day, and feeling much less nervous. She talked naturally, unless unusually excited, and controlled her limbs perfectly. She was allowed to return home, with instructions not to begin any eyework for at least three months.

The change in the facial expression of this patient, as well as in her walking, talking, and general demeanor, that had occurred in less than six weeks was so marked as to excite the greatest surprise among those who had met her when suffering from her nervous malady.

Although I did not regard the latent esophoria in this

case as fully corrected, yet the physical condition of the patient had so improved that I deemed it best to wait several months before any further investigation of her remaining heterophoria. Subsequently this patient was taken violently ill at her distant home and has never returned to complete her eye-treatment.

CASE XIII. *Frightful Contortions of Face and Incessant Jerking of Head for Two Years.*—Mr. D., aged 42; married; contractor.

Family History : Father living, aged 80. Mother died of cancer. One sister died of phthisis. One sister has severe headaches. One child, aged 12, has sick headaches. One brother died of paresis. One brother died of mental trouble "due to sun-stroke."

History of the Case : This patient had suffered from headaches, more or less frequently, all of his life, but for the past two years they have been practically constant. One day out of three the headache is liable to be very severe. The pain is increased by overwork, reading, or excitement.

Two years ago he developed, very suddenly, an attack of dizziness. He then weighed two hundred pounds. Chorea gradually developed until it became extreme. At no time has there been any paralysis or loss of power. For over eighteen months he has been jerking violently in his face, arms, and legs. He has consulted many physicians, without any marked improvement. The jerking movements are more violent in the forenoon than later in the day, and more in cold weather than in warm. He has lost fifty-five pounds in weight. He sleeps and eats well. He has to use cathartics and injections to overcome extreme constipation.

This patient was a very prominent man in political circles. He had spent large sums of money in consulting numerous physicians, and his case had been a great puzzle to all the medical men who had examined him. One of his friends sent a communication to me, before I had treated the patient, that "if anything could be done for this patient by eye-treatment it would be a miracle."

This *patient had had no eye-symptoms.* The jerking of his facial muscles and his head was so extreme as to make it very difficult to make satisfactory tests of his eye-muscles,—in fact, his head had to be held firmly in the grasp of an assistant before any tests could be made.

Eye-tests : Prior to the use of atropine this patient showed an apparent myopia of 1.25 in the right eye; emmetropia in left eye; no hyperphoria; esophoria, 2°; adduction, 17°; abduction, 6°. After the instillation of homatropine the following tests were observed: O. D. V., $\frac{20}{40}$, made $\frac{20}{30}$ by — 0.50° $\bigcirc$ — 0.50° (axis, 180°); O. S. V., $\frac{20}{30}$, made $\frac{20}{20}$ by + 0.50°. Under the judicious use of prisms this patient disclosed a high degree of latent esophoria with homonymous diplopia, and no abduction.

Treatment and Results : This patient was ordered a full correction for his refractive errors for constant use. A graduated tenotomy was done upon his right internal rectus, the immediate results of which were that he showed exophoria of 1° with abduction of 11°. Later he showed orthophoria.

Within twenty-four hours there was a wonderful decrease in his jerking movements. Eye-tests could be made upon the patient without holding his head. Four days after the operation patient reported that he had taken no cathartic and his bowels were perfectly regular. Within a month he was almost free from chorea.

For *nearly three years this patient has been absolutely well.* No further operative work has been required upon his eye-muscles. He is transacting his large business without ill effects upon his health. The removal of his glasses causes him immediate discomfort.

CASE XIV. *Terrific Jerking of the Head and Face, with only Slight Spasmodic Movements of the Arms and Legs.*—Mr. S., 34 years old ; single ; cigar-manufacturer.

Family History : The patient seemed to be very uncertain regarding his ancestry and the diseases that had developed among them. Mother died of paralysis. Sisters have headaches. No phthisis.

History of the Case : This patient was a cigar-maker that lived in one of the cities in the western part of New York State. He had been perfectly well up to six months prior to his visit to my office. Fifteen years ago he had a venereal sore, followed by suppurating glands in the groin. He has never had any eruptions of the skin, sore throat, or falling of the hair.

Six months ago he began to have a twitching in the muscles of the neck and pain in the back, between the shoulders.

This condition grew rapidly worse until he had terrific jerkings of the muscles of the neck and face, with a pain running down the spine as far as the lumbar region. The arms and legs were only occasionally affected to a slight degree.

As the possibility of syphilis was suspected, he was treated, for awhile before I saw him, with iodides and mercurials. No improvement followed, and syphilitic treatment was abandoned. He then underwent a course of electrical and massage treatment, without benefit. Arsenic and all other therapeutic remedies proved of no value in his case.

When sitting down or remaining quiet his convulsive movements were only moderate in severity, but when he attempted to walk the distortions and grimaces of his face and the violent jerkings of his head were distressing to witness.

As he was considered a very remarkable case by the doctors whom he had consulted, and as no improvement had followed their efforts, he was referred to me for diagnosis and treatment. He stated to me, on the first interview, that the only improvement from the various forms of treatment which he had tried was that he could walk one short block, sometimes, without a violent attack of jerking. On arrival in New York City he obtained lodging in a hotel two blocks from my office. On the second morning, while coming to my office, he was seized with one of his paroxysms on the street and was arrested by a policeman on account of the grimaces of his face and the violent jerking of his face, head, and neck.

Eye-tests : Patient had normal vision in each eye, but tolerated a weak spherical glass ($+ 0.50^s$). He showed only : Esophoria, $\frac{1}{2}°$; no hyperphoria ; adduction. $32°$; abduction, $8°$. After the instillation of homatropine he showed, in each eye, latent hypermetropia of 2.50 diopters.

In spite of his high abduction I suspected the existence of latent esophoria. Under the judicious use of prisms he soon disclosed (while wearing $6°$ of prism) esophoria of $8°$ with abduction of $2°$.

Treatment and Results : This patient had not sufficient means to remain in New York for treatment. I therefore *ordered for him* $+ 1.50^s$, $\bigcirc 3°$ *prism, base out, over each eye,* and I instructed him to wear his spectacles constantly. I advised him, also, to consult a local oculist at his home, from time to time, who had been personally under my care and who could report to me the eye-tests of the patient as often as he deemed it necessary to do so.

Within two weeks the spasmodic movements had practically ceased.

For nearly nine months no return of his terrible affliction has occurred. He was wearing the same glass as originally prescribed, although, in my judgment, the spherical glass ought to have been increased and the prisms discarded, after a radical relief of his esophoria by a tenotomy upon an internus.

He then was persuaded by a local oculist to have both interni operated upon. Plain spherical glasses ($+$ 1.50) without prisms were then ordered for constant use.

This patient called at my office on April 15, 1897, and reported that the cure was complete. He showed not even a trace of facial movement and his eye-tests were practically normal so far as heterophoria was concerned. The patient said that his recovery was regarded, by all who had known him when afflicted, as miraculous.

CHAPTER V.

SLEEPLESSNESS: SOME FACTS RELATING TO ITS CAUSES AND CURE.

A PERSISTENT loss of sleep is generally recognized, and properly so, as one of the most dreadful of human afflictions.

The recuperation of vital forces that takes place during peaceful slumber constitutes the basis of both mental and physical health. "Tired nature's sweet restorer—balmy sleep"— comes, from natural causes, as an unbidden, but welcome, guest to the many that would otherwise be unable to bear the burdens of each day.

To those who are robbed of sleep, however, from causes that may appear obscure, the struggle of life sooner or later ends in physical or mental disease. It entails upon them too often the distressing results of the opium or chloral habit, and, in many instances, a sudden termination of their misery by suicide.

There is probably no physical condition that the educated physician is so often unable to combat successfully as a persistent tendency toward insomnia.

Many thousands of sufferers of this class are to-day exiled, through medical advice, from their business interests,—traveling for needed rest or enduring a wretched existence in uncongenial surroundings, far from their homes and friends, too often at sacrifices that they can ill afford to bear,—simply because medicine, in their special case, has proved incapable of combating an inability to obtain the eight hours of peaceful slumber that health demands.

Again, it is chiefly among the intelligent and educated classes that this form of suffering is encountered.

Those by whom, in consequence of their mental endowments, the greatest successes might be, and often have been, achieved; by whom the greatest pleasure can be obtained out of life and alike bestowed upon others by their mental and social prominence; upon whom the greatest responsibilities are of necessity imposed, and in whom the greatest capabilities of suffering and patient enduring exist—these are too often the

(136)

ones that turn, in their despair, to the physician for relief from this dread enemy that shatters hope, paralyzes industry, impairs the judgment, imperils large financial enterprises, embitters life, and casts a gloom over their present and their future.

The explanation of the fact that brain-workers (as distinguished from muscle-workers) are peculiarly disposed to insomnia lies chiefly in the constant strain imposed upon the organ of the mind; but this does not necessarily justify the conclusion (too frequently arrived at by medical men) that the brain-cells, or the blood-vessels that feed the brain-cells, are the seat of actual disease.

At the risk of incurring criticism for repetition, I feel that I cannot impress my readers too frequently with the physiological fact that any expenditure of nervous energy in excess of that generated from day to day (irrespective of where the excessive expenditure occurs) may in time so deplete the reserve-capital of nerve-force in any individual as to embarrass the workings of some part or parts of the nervous system without any actual disease being present.[1] The result of this temporary "nervous bankruptcy" is peculiarly apt to disclose itself in some derangement of the normal function of the weakest part,—as an echo is heard far from the source of the echo.

Let us cite, as an apt illustration of what I mean, one of our every-day experiences :—

An upright businessman, with a stated income, has, from certain extravagances, etc., spent not only in excess of his income for many years, but has gradually encroached upon his capital. He grows moody, reticent, and irascible; and becomes, almost imperceptibly, an altered man. His friends, ignorant of the cause of the change, gradually become distant and fewer in number. Social estrangements, domestic unhappiness, a general loss of esteem, and many other complications then begin to arise, day by day and month by month, until the individual falls from the high position that he once occupied with warrantable pride. Now, what has caused this fall, and what is the remedy? Unquestionably, to every thinking mind, the initial and underlying factor in all the ultimate results would be the excessive expenditure of money. The cure, moreover, lies in stopping the initial cause, with the hope that time and prudent

---

[1] The closing pages of a preceding section (pages 16 and 17) cover this argument in a more exhaustive manner.

living will restore not only the impaired business capital, but
likewise the cheery nature and honest manhood that originally
gained the individual his high position, and that can alone
restore it to him.

The reader may possibly fail, at first, to see the application
of this illustration to the subject at issue. He will see it in a
clearer light when I submit the following proposition, which I
shall endeavor to sustain by a report of cases that have come
under my observation. This proposition I would present as
follows :—

*Clinical evidence goes to show that a large proportion of
subjects affected with persistent insomnia of long standing suffer
from some congenital defect of the eyes themselves or from an
improper adjustment of the muscles that move the eyes.*

In many incurable cases of insomnia this constitutes the
underlying factor that entails an excess of nervous expenditure
from the date of birth until death (if not properly rectified).
In time it materially tends to exhaust the normal reserve-capital
of nervous force of the individual.

When we stop to reflect, we can understand how every
letter on a printed page, as well as every object on the street or
in our homes that we become cognizant of by the sense of
sight, requires a more or less perfect adjustment of the complicated muscular apparatus that so regulates the eyes in relation
to each other as to enable them to see with both and yet perceive but a single image.

The total *aggregate of such visual perceptions*, during the
sixteen hours of each day that we use the eyes, is enormous ;
and it means a proportionate number of accurately-performed
adjustments of two cameras (the eyes) upon a single object,
performed often with marvelous rapidity and involving, in many
of the adjustments, a complete change of combinations in the
eye-muscles that are successively brought into play. It is not
much of a task to lift a penny once, but no living being could
lift a penny a million times each day.

Now, nature has so accurately balanced the relative power
of each of the various eye-muscles in a perfectly-constructed
being, and has so beautifully constructed the eyes as regards
their focus, that the expenditure of nerve-power (in the case of
such an individual) required to perform the necessary eye-movements throughout each day is reduced to a minimum, although

necessarily very large as compared to the amount expended upon any other organ in the body.

But, when the adjustment of the eye-muscles or the construction of the eyes themselves is so imperfect that the maintenance of single vision (when both eyes are simultaneously used) is the *result of an excessive expenditure of nerve-force* (far greater than nature intended in many cases), any individual so afflicted begins from birth either to draw from the " reserve-capital of nerve-force " that nature has stored up for emergencies, or the eyes must be run at the expense of a proper nerve-supply to some other part (Peter being robbed to pay Paul).

As I have said before, in a preceding chapter, three factors enter into the proposition as to how long a time can elapse before the serious influences of such a leak of nervous energy will be felt in any given case where the eyes or the eye-muscles are abnormal: 1. How much excess of energy over the normal amount is required to compensate for the defects connected with the sense of sight. 2. How much " reserve-capital " of nerve-force the individual starts out in life with. 3. How much nerve-force the individual can generate day by day to meet the daily expenditure.

A person inheriting one hundred thousand dollars at birth could have expended upon him one thousand dollars per year in excess of his income without feeling the lack of money for one hundred years ; but if the excess of expenditure be increased to five thousand dollars over his income, bankruptcy would stare him in the face when he attained his majority.

A serious defect of construction in one or both eyes, or a decided tendency of one or both eyes to deviate from parallelism with its fellow, may entail upon an individual a leakage of nervous force that is apt to produce, in time, very sad results upon the general health.

It is not my intention in this article to ignore the fact that many cases of sleeplessness can apparently be attributed to overwork, business cares, anxiety, or similar forms of nervous strain as one of the factors in its causation. Neither would I overlook the fact that organic disease of the kidneys and diabetes often manifest their onset by a persistent tendency toward wakefulness. I desire simply to emphasize the fact that eye-strain constitutes (in a large proportion of such cases) a factor that is often unrecognized or ignored by medical men.

The field of the oculist is of necessity a large one. It will not be restricted by a wider dissemination of knowledge of the apparatus of sight among the body of general practitioners.

Within the next twenty years we will see every specialist of note in nervous diseases more or less expert in testing refraction for himself, and determining, without outside aid, the existence of defective equilibrium of eye-muscles. He need not be an oculist, in the true acceptation of that term. Diseases of the eye properly belong to a specialty, and are best treated by those who see the most patients of that class. But the nervous specialist should know, and sooner or later will be forced to know, whether his patient has near-sightedness, far-sightedness, or astigmatism to complicate matters, and if the eyes tend to deviate from their normal and physiological conditions. The more he studies these conditions, the more will he find to interest him, and much to relieve that is now too often unrecognized in the sufferers who apply to him for aid. He will cease to give pills to patients with from three to twenty poisonous alkaloids combined, on the principle that the sportsman adopts when he uses a handful of shot, hoping that one may perchance kill his bird. He will study his cases more intelligently, and delve less into works of therapeutic speculation. This view is not Utopian, or visionary; neither is it the result of rash enthusiasm; but is an earnest conviction that has come after years of patient inquiry and careful observation in a large number of patients suffering from nervous derangements.

The time has passed, I think, when the blindness of prejudice against the views advocated in this work holds sway as strongly as in the past among the leading minds of the medical profession. From many sources, both here and abroad, abundant confirmatory evidence of the truth of this doctrine is being published from time to time, showing conclusively that defects of the eyes and eye-muscles do constitute an important factor in the causation of many forms of obscure nervous diseases. Those who still oppose this view most vehemently have, in many instances, shown by their writings gross ignorance of the methods employed to decide the points at issue.

It took Lister and his followers some ten years to teach the profession, if the antiseptic method was to be tested as a basis for adverse criticism or for the benefit of suffering humanity, that the operator must clean his finger-nails; that he must also wash his hands with great care, make his knives aseptic, and follow out the published plan of procedure with due regard to detail before the results obtained could be worthy of publication or in any way reliable as a basis of scientific deduction.

To what extent the leaders of the medical fraternity willingly lend their ears and give their earnest support to any new method of treatment [provided it consists of a drug, or a subcutaneous injection of an agent whose component ingredients are unknown and which is adopted purely on faith] has been demonstrated during the past few years in a way that now bids fair to subject such child-like credulity to ridicule,[1] if the ill results to humanity do not in time justify merited condemnation and rebuke.

On the other hand, any system of treatment that is based upon facts which can be determined with the same scientific precision as the computation of an astronomical problem, that is supported by well-recognized physiological laws, that has yielded and is yielding daily relief to many individual cases culled from suffering humanity which medicines have failed to afford, that has lived and steadily made progress year by year in the face of bitter and organized opposition—such a system of treatment cannot now be annihilated by ridicule as a substitute for scientific argument or impeded in its progress by the condemnation of those who have had no experience in it.

Not long ago a famous orator told the following fable:—

" A well-fed horse who, in his greed, scattered grain upon the floor of his stall became, in consequence, the constant companion of a rooster, who picked up the scattered oats. One day the rooster suggested that friendly relations were desirable and would be put on a much firmer basis by the existence of a solemn agreement between them. The horse assented, and, on asking the basis of the compact, was told that it should read: *Neither of us shall step on the other's feet.*'"

I do not desire to carry out in full what, to my mind, might be the true application of this fable. None of us desires to stir up discord if important facts can be insured a fair hearing without recourse to asperity; but the establishment of a great truth cannot be crushed by being " stepped upon."

---

[1] I refer to the injections subcutaneously of certain animal extracts that are to-day practically abandoned.

I do not purpose to discuss here at any length the optical problems (often very difficult to solve) which are liable to be encountered in subjects who suffer from serious nervous disturbances of the functional type. Neither is it my intention to review here the work that has already been done in this field by the employment of methods now in vogue, old or new.

It may be well, however, for me to mention, in this connection, a few of the reasons why, in my judgment, the treatment of the eyes has totally failed, in the hands of some observers, to relieve or modify some nervous conditions that had withstood judicious medication for years, and why it is that, subsequently, in more experienced hands, treatment of the same patients, directed to the eye-muscles, has led not infrequently to the happiest results.

1. I would call attention to the fact that *preconceived notions about old methods must be abandoned without prejudice* when a new method is to be tried.

2. Each observer must, of necessity, make himself *thoroughly familiar with all the details of the method* which he purposes to employ before he is competent to decide *pro* or *con* respecting its merits. This cannot be done exclusively by reading. No one can describe with the pen the many intricacies that are apt to arise in solving complex optical problems. It is certainly not beneath the dignity of even an eminent man to learn (by personal observation of the work of another whom he, perhaps, thinks is misled, and by timely suggestions thus obtained) how facts that bear upon successful treatment may be determined that were, perhaps, at first obscure and difficult to ascertain.

3. With a full knowledge of the method, its intricacies, and its difficulties, *conclusions should never be too hastily arrived at* in any given case. It is always "better to be sure than sorry." Those who have had the largest experience may occasionally make mistakes in judgment when a peculiarly complex problem is presented for solution. How much easier is it, therefore, for one with a limited experience to fall into error! The story is told that a selection of a pilot for a vessel laden with precious merchandise, which was to enter a harbor full of sunken ledges and sand-bars, was once being made. One by one the applicants told a tale of uninterrupted successes. Finally one pilot was accepted simply because he said: "I ought

to know the channel, as I've wrecked a ship on every rock in this harbor."

So it is with many cases of epilepsy, chorea, insanity, insomnia, neuralgia, headaches, and kindred nervous affections. These patients have, as a rule, acquired and constantly practiced from birth *certain faulty combinations of the various eye-muscles*, in order to enable them to use the eyes together.

They are often able, by the aid of such unnatural combinations, to *simulate* a condition of apparent equilibrium of adjustment of the eyes, although a very serious expenditure of nerve-force may be demanded of them in order to do so. They are naturally unconscious of the eye-strain, because they think everybody does as they do, in order to see. They often have no eye-symptoms. They practice these "tricks of adjustment" instinctively, not as an act of volition; and they have generally to be taught, by the aid of prismatic glasses and other recognized steps, to abandon them and thus to disclose the actual maladjustment of their eye-muscles, which has entailed upon them this long-continued leak of nerve-force, and, later on, an abnormal reflex excitability of the nerve-centers.

They are not unlike a tortuous and difficult channel in which the hidden difficulties to surmount do not always disclose themselves upon the surface. They demand, and generally repay, a careful and scientific scrutiny into the adjustment of the ocular muscles and latent errors in refraction. They require time and great patience on the part of the observer, as well as skill.

4. The old methods of testing the eye-muscles will have to be abandoned at no distant date. A phorometer is now essential to all accurate work. Moreover, the separate muscles should be individually tested and their power accurately measured.

Not long since, a physician, who had twice collapsed from nervous prostration and insomnia at the very threshold of his professional labors, came to me for advice. He showed, at intervals, an apparent condition of equilibrium in the orbits, but welcomed prisms for a deviating tendency of one eye above its fellow, and improved rapidly under their influence. Within a week he showed unconquerable double images without his prism, and a radical step for the correction of his vertical strabismus was advised. At the advice of friends he then consulted an oculist of international repute, who not only failed to recognize the fact that the patient saw double images, but even pronounced the eyes normal in their adjustment. The description by the patient of the rough and unscientific tests upon which that judgment was made showed clearly that the oculist was either woefully negligent of his obligations to the patient or incompetent to decide the point at issue.

Another patient, upon whom I have lately performed a graduated tenotomy of the external rectus muscle, with the happiest results (as it brought about a rapid and complete restoration to health), came to me originally with an eye that diverged, at times, when her vision was not attentively engaged, almost to the outer canthus; yet she bore a certificate from one of the leading oculists of America that she had no defect in the refraction or adjustment of the eyes and that her terrible headaches and difficulty in using her eyes required only constitutional treatment.

Hardly a day now passes that I do not receive, from some patient, written testimonials or verbal statements of the deepest gratitude for relief that has come to them through the correction of some defect in the apparatus of vision (often unsuspected by the patient); and the intensity of their expressions of gratitude is unquestionably based, in many of these patients, upon the fact that drugs had been administered to them for years, according to the latest therapeutic theories, without any perceptible benefits.

One of my warmest professional friends, and an eminent medical teacher,—Dr. J., whose wife has been relieved of insomnia and some other nervous symptoms of some years' standing by the aid of glasses alone,—writes me, in the exuberance of his gratitude, as follows: " Your name will occupy a prominent place in our shrine, and heaven will be continually besieged to bless both you and yours." (See record of Case IV of this chapter.)

Another patient, at one time a hopeless victim to insomnia, writes me, after an operation on an eye-muscle and the fitting of proper glasses: " I cannot trust myself to think of what the final result might have been had you left me to my own will last August. I have had what I cannot pay for in money, even if I could send you my check for a very large amount."

Dr. R., an eminent divine of this city, says: " I am better

since your treatment than ever before in my life, in my sleep, digestion, pulse, calmness, vigor, and eyes."

It has been denied most emphatically in the past by some specialists of prominence, both in eye diseases and nervous affections, that the eyes or the eye-muscles have any marked influence upon nervous diseases.[1] Statements of this kind are still made by some, both verbally and in print, with the same vehemence (based, it is to be feared, upon bitter prejudice and partisan feeling) as they were years ago. Despite evidence to the contrary, they refuse to see what many non-partisan minds are seeing more and more clearly every day. They discard, without trial, new methods of research into complex ocular problems; they prohibit the use of instruments (that alone allow of an approach to scientific precision) from institutions that they control; and in every possible way they appear to try to mislead the line of professional thought from the main points at issue.

One thing is evident,—viz., the view that eye-strain can and frequently does cause serious nervous conditions must be either true or false.

If it be false, then it has made steady progress in spite of its weakness and against organized and bitter opposition; if false, then the growing list of converted advocates among the younger oculists and neurologists is incapable of explanation; if false, then the thousands of suffering humanity are deceived who believe that they have cause for the deepest gratitude in the recognition and relief of an existing eye-strain. It is contrary to all precedent that a mere " fad " should steadily flourish and gain strength year by year over a period of many years; neither does the statement that some cases have failed to be benefited by this treatment have any weight in argument.

---

[1] It may not, perhaps, be best to quote here the words themselves or even the titles of articles that have been published in the past by medical men of note condemning these views. Some would, I know, like to withdraw to-day what they foolishly said and wrote some years ago, when the methods advocated here were in their infancy.

I cannot refrain, however, from stating in this connection that I listened a few years ago to a most vehement harangue by an oculist of note before a medical assemblage. It was filled with absurd statements delivered in *ex-cathedra* fashion; ended with a peroration of denunciatory comments and conclusions; and was based *in toto* upon the most manifest ignorance of the first principles of the method that he condemned. Within a week I heard this same oculist confess: "I have never tested an eye-muscle (presumably by the methods he had condemned or any other); I never shall do so while I live; and there is nothing in it."

Does not patience cease to be a virtue under such circumstances? Is it not a sad commentary upon the value of some medical opinions, when bigotry and prejudice run riot?

Every method of treatment of disease sometimes fails to relieve individual cases; yet no one attempts to discard all therapeutic efforts in consequence of this fact, because such a deduction would be manifestly illogical.

The following cases illustrated the remedial effects of a relief from eye-strain upon sleeplessness as a prominent symptom :—

CASE 1.—Miss B., aged 40 ; lecturer.

Family History : One sister was for over a year a victim to " complete nervous prostration." Father is a very nervous man.

Eye-defects : Vision. $\frac{2.0}{5.0}$, without atropine ; under atropine, a latent hypermetropia of $+ 0.75^s$ in each eye ; patient had never used a glass for reading, but accepted $+ 1.50$ spherical glass ; esophoria, 3° (which ultimately, under influence of prismatic glasses, exceeded 7°) ; adduction, 23° ; abduction, 5° ; right sursumduction, 1° $+$; left sursumduction, 2°. The adducting power later on exceeded 43°, and the abducting power fell below 3°. At no time did homonymous diplopia disclose itself (with or without a red glass).

History of the Case : This lady had for some years been doing an excessive amount of mental work. Her profession required an enormous amount of reading. This had been done largely at night. Although small in stature, she had always been vigorous and had taken an unusual amount of exercise. She *had always considered her eyes very strong.* and was loath to believe, when she first came under my care, that her eyes could constitute a factor in her serious nervous condition. Furthermore, she was strengthened in this belief by the fact that she had not long before consulted an oculist of prominence, who had stated that he found no defect requiring treatment or glasses, and who had sent her to one of his friends (a specialist in nervous diseases) for treatment.

The " breakdown in her health " began about twelve months before she came under my care. It was attended with an *extreme and persistent loss of sleep,* a loss of emotional control, an utter inability to read or sew (which aggravated all her symptoms), a more or less constant headache. an inability to concentrate her intellectual faculties for any length of time, and an aggravated type of mental depression.

She feared, and had every apparent reason to fear, that her

professional labors were imperiled and that her mind might possibly give way. The neurologist, who endeavored to build her up by tonics, rigid diet, rest, etc., assured her (after some improvement had occurred) that he feared at first that " melancholia " might be the end of the case. At his advice, she spent the summer at the sea-shore; but, beyond a certain point, she failed to progress satisfactorily, and her headache and sleeplessness would, at times, be as bad as ever. Any attempt to prepare herself for her fall engagements would cause a return of her old symptoms to a very marked degree, accompanied by physical weakness, mental fatigue and depression, extreme despondency, and a lack of control over her emotions. After any attempts at study she would frequently lie awake most of the night. This was her condition when she first came under my care.

Treatment and Results: In this case a full correction of the hypermetropia was made for distance, and + 2.00 spherical glasses were given for reading, as she showed some failure of accommodation. Prisms of various strengths were employed over her distance and reading-glasses for about two weeks, and 7° of latent esophoria were found to exist. This was rectified by a graduated tenotomy of one internus and the prisms were then discontinued. During this interval the patient had improved very rapidly, had become very dependent upon her spherical glasses, and become cheerful and hopeful of recovery. She had, moreover, entirely regained the normal power of sleep. During this interval she had frequently slept twelve hours without awakening and without recourse to any drug. As atropine had been used during the early part of the treatment, she had been allowed during the two weeks of treatment to use her eyes very little in reading or study. During the following two weeks 2° more of latent esophoria disclosed itself. For the relief of this defect a prism was combined with the spherical glass worn over the eye which had not been subjected to a tenotomy.

For the next five months this patient was able to fill all her engagements without any return of her bad symptoms. She had read and studied at night, attended church and places of amusement that previously she dared not attend, had accepted more work than for some years past, and had continued to sleep well and enjoy perfect health. During this interval she took no medicine, nor had she been restricted by me in her diet or in any other way. Her reading-glasses had been increased to + 2.50ˢ.

The records of her case show that a graduated tenotomy of the internus of both eyes would eventually have to be performed, in order to properly adjust the balance between the two eyes.

During one of her last visits this patient said: "I think I am stronger to-day and have better health than I have had for many years. I certainly do my work with less fatigue, and enjoy things that my ill health has previously debarred me from."

Some months after this visit this patient died from an accident.

CASE II.—W. C., aged 42; minister of the gospel; married.

Family History: Mother living (aged 84). Father was a delicate man and had headaches. One brother and one sister have severe headaches. One brother has several times broken down in his studies from eye-pain and asthenopia.

Eye-defects: Myopic astigmatism, O. D., — 1.25°; O. S., — 0.50°; exophoria, 10°; crossed diplopia, 4° (with red glass before either eye); no hyperphoria; adduction, 18°; abduction, 12°; right sursumduction, 3°; left sursumduction, 3°; presbyopia (uses + 1.00^s for reading).

History of the Case: This patient was never a very strong man and had always been a hard student. About six years ago he began *to suffer from sleeplessness,* confusion of thought, and an utter inability to apply himself to his work for more than a few minutes at a time.

He put himself in the care of a prominent physician of this city, and at that time was examined by an oculist of note, who prescribed glasses to correct his astigmatism. These gave some temporary relief, but he soon broke down completely, resigned his position, and for many months was unable to do any work. After a long rest he again took charge of a church, but his insomnia soon became so severe that an extended vacation was necessary. Since that time, by the most regular habits, careful diet, and daily exercise in the open air, he had been able to keep his position, although he feared a collapse at any moment. Finally, his loss of sleep, distress in his head (although he had no actual headache), and confusion of thought became so constant that he was again on the verge of resigning his position when he came to me for treatment.

Treatment and Results: The treatment of this case consisted simply of two graduated tenotomies for the relief of the exophoria, and subsequently a reading-glass to correct his presbyopia. The result of this treatment was most gratifying. A report received from the patient within three weeks after the first tenotomy states that he "now sleeps well and has great relief from the distress which he has so long experienced in his head."

For over four years, since his eyes were treated, he has been doing his work with more comfort than for many years. He sleeps well, with the exception of an occasional poor night after an unusually trying day's work. He has taken no medicinal treatment.

Case III.—A. C. H., aged 46; manufacturer; married.

Family History: Both parents lived to 76 years. Two paternal uncles died of phthisis. No hereditary tendency to nervous diseases.

Eye-defects: Vision, $\frac{20}{30}$, without atropine; under atropine a latent hypermetropia of $+ 1.00^s$ in each eye; patient had never used a glass for reading; esophoria, 5° (after using prismatic glasses for a short time the patient showed esophoria of 13°); adduction, 24°; abduction, 4° +. Later on the adduction exceeded 50° and the abduction fell to 0. Homonymous diplopia with the red glass over one eye was usually present, and at times without the red glass.

History of the Case: This patient had been a perfectly well man and had carried on a very large business up to fifteen years ago. At this time, while attending a sale in New York, he was suddenly seized with a dizziness, faintness, and a sore feeling in his head.

These symptoms continued for three years in spite of all treatment, during which time he *suffered severely from sleeplessness*, extreme nervousness, and soreness in his head. He was unable to look out of a car-window, while traveling, without great distress.

He had suffered all his life from obstinate constipation, and had taken cathartics so regularly that now any cathartic water causes intestinal hæmorrhage.

When this patient first came to me he was able, by the most careful diet, regular habits, and by retiring at eight or nine o'clock. to carry on his enormous business only with the greatest difficulty because of the following symptoms: Inability to sleep at night, which at times was very distressing and persistent; extreme nervousness after the slightest fatigue; mental depression without any cause; hot flashes up and down his spine; pain in his shoulders and across his back. His insomnia was often prolonged and very exhausting after any slight excitement or fatigue.

Treatment and Results: The treatment of this patient consisted at first of the wearing of prisms to relieve the esophoria, and later on of graduated tenotomies on both internal recti. Subsequently, + 0.50° glasses were given for constant wear and + 1.00° glasses for near work. The improvement in his condition was marked and continuous from the first, and he writes that he is so busy and feeling so well that he cannot find time to have the slight remaining esophoria corrected. An extract from a letter received from him six months after the operation on his eyes speaks for itself. He says: "Seemingly I am all right. feeling better every day; have not had a headache for a month; appetite good and I sleep well." Over five years have now elapsed without any return of his former ill health. during which time he has constantly been engaged in active business pursuits.

Case IV.—Mrs. J., aged 45.

Family history not taken.

Eye-defects: Hypermetropia and astigmatism of + 1.50° ◯ + 0.50° in each eye (under atropine); right hyperphoria, ¼°; esophoria, 0 to 1°; adduction, 21°; abduction, 8°; right sursumduction, 2° +; left sursumduction, 2°.

History of the Case: For many years patient has been a delicate woman, becoming easily fatigued, and suffering more or less after fatigue from insomnia and extreme nervous debility. For the past ten or twelve years one pupil has been very much dilated. She had consulted an oculist of prominence in Montreal concerning this condition. but his treatment failed to give any permanent benefit. During the past twelve months the insomnia and nervous prostration had become very much intensified. and the patient had become so weak physically as to alarm her family. Any attempt at walking, attending places of amusement, or making ordinary social visits was followed by a marked increase in the symptoms. Her husband, a prominent physician, feared a complete physical collapse. One pupil was found to be more than double the size of the other.

Treatment and Results: The treatment consisted of a full correction of the hypermetropia and astigmatism by glasses, which the patient was instructed to wear constantly. Under these conditions her muscular tests seem to be modified favorably. The patient was instructed to return home and to come again for further observation after wearing the glasses for a few months. Even before her return there had been a marked improvement in her symptoms. Two weeks afterward the following report was made by her husband: " My wife appears much better and more cheerful than for many years, the pupils are of equal size, appetite good, and the insomnia much relieved; is able to walk two miles without fatigue and enjoys the exercise, goes out evenings, and feels no unusual fatigue from lectures, concerts, and sermons." A report some months later says: " My wife appears to enjoy life as she has not done for many years. There has been a very slight return of her old enemy, insomnia, but not to an alarming extent. She hopes to see you again in the near future."

In this case sufficient opportunity has not yet been afforded for a complete examination of the eye-muscles. It is possible that there may be some lurking defect of equilibrium in addition to the error of focus. One thing, however, appears to be clearly established,—*i.e.*, that her ill health and insomnia were directly dependent upon a condition of the eyes that had exhausted her vital forces and was keeping her in a state of extreme physical depression.

CASE V.—Mrs. W., aged 55.

Family history not taken.

Eye-defects: Hypermetropia, $+ 1.75^s$; presbyopia (uses $+ 4.50^s$ for reading); esophoria, 7°; adduction, 23°; abduction, $3^\circ +$. Later on she disclosed: Right hyperphoria, 3°; right sursumduction, $6^\circ +$; left sursumduction, $2^\circ -$.

History of the Case: This patient is the wife of a prominent physician, and, as such, has had the benefit of the best medical talent of the State in which she resides. She had always been a delicate woman up to the time when my professional opinion of the case was asked.

For a year or more before I first saw her she had been a victim to nervous prostration and confined most of the time to her bed or room. Her life had been despaired of during this interval at times, and the case seemed to present problems in diagnosis which puzzled the best medical men whom she had consulted. When she had gained sufficient strength to allow of her being moved with safety, her husband was advised to take her to a Southern climate. On their way to Florida he was advised to consult me, in reference to the case, when he passed through New York.

When I first saw this patient she was in a state of extreme physical and mental depression, was unable to walk for even short distances without great fatigue, was sleepless and despondent, and was brought to my office in a carriage from a hotel not far from my residence.

Treatment and Results: At the first visit prisms were given to relieve the esophoria, and in five days a graduated tenotomy was done on one internal rectus. The patient began to feel the benefit of this step from the first. The second day after the tenotomy she reported that she had walked a mile and a half,—a thing which she had not done for over a year. Five days after the first tenotomy a second one was performed on the other internal rectus, prisms having been worn in the meantime. Two days following this the patient walked five miles, visited an art-museum in the morning, and attended a theater in the evening. In spite of the unusual fatigue and excitement she was still sleeping well and feeling stronger than for many years. With the improvement of her general health came an entire cessation of her obstinate bladder-trouble, which had given her annoyance for many years and was probably due to her weak muscular and nervous condition. The pain in the bladder, which was probably of the neuralgic type, ceased after the relief of the eye-tension, and has never returned.

After an interval of four months, during which she had been comparatively well, she returned to New York to complete her treatment. A high degree of hyperphoria was found, and prisms were combined with her hypermetropic glasses to relieve it. With these glasses the patient passed eight months of almost absolute freedom from distress of any kind, when a graduated tenotomy was performed and the hyperphorial prisms removed.

At the present time she is sleeping well, is able to attend to her household duties. can walk long distances, has taken no medicine for over five years, and is regarded by her husband and friends as restored to perfect health.

CASE VI.—Mr. P., aged 41; manufacturer; married.

Family History : Both parents died at 75 years of age. Mother was of nervous temperament. One brother died of Bright's disease.

Patient has seven children. all unusually healthy.

Eye-defects : Hypermetropia and astigmatism ; O. D., + 0.50^s ◯ + 0.75^c (axis, 90°) ; O. S., + 1.25^s ◯ + 0.50^c (axis, 90°) ; left hyperphoria, 4° ; esophoria, 7° ; adduction. 22° ; abduction, 4° ; right sursumduction. 2° ; left sursumduction, 8°.

History of the Case : This patient was brought to me by his wife, from Canada, at the suggestion of their physician, Dr. B., who had good cause to suspect an advanced case of soften-ing of the brain.

For several months prior to this visit the patient had taken but little, if any, medicine and had steadily lost flesh. His mental condition had become alarming, and his doctors had practically regarded the case as hopeless. He had to be taken care of by his wife, who paid all the bills and looked after him as she would a child. While dressing he had to be told what clothes to wear and which to put on first. At the table he would chew his food until told to swallow it. His demeanor was extremely apathetic, except at intervals. when he would start suddenly from his chair, grab his head with both hands. and walk in an agitated manner about the room, complaining of great pain in his head.

For twenty years he had had severe attacks of neuralgia affecting the left eye and left side of the face, and for many years he had been annoyed by flowing of tears over the cheek in cold weather. For six years he had had marked symptoms of indigestion, flatulence, and constipation. Eating was followed by severe pain in the region of the stomach. and he had been obliged to restrict his diet for some years in consequence.

For six months prior to this visit he had not visited his place of business and *had suffered terribly with insomnia.* To

such an extent did the insomnia exist that his wife would sit by
him and fan him during cat-naps until noon of each day.

As he was not in condition to stand the excitement of a
hotel, he was forced to lodge with friends where absolute quiet
could be insured.

Great difficulty was experienced, during my first few interviews with the patient, in getting any reliable eye-tests, although
his expression indicated a marked degree of left hyperphoria
and esophoria.

After reading only a couple of lines on a test-card he would
leap from his chair, grasp his head with his hands, and say that
he would come in the next day and read some more.

The case certainly looked most unpromising, and his mental
condition was such that I could not divest myself of the belief
that organic brain disease existed and that the case was probably
incurable.

After several interviews and the free use of atropine to
dilate the pupils, the eye-tests became more satisfactory. I
advised the wife to consider the propriety of an operation for
the manifest hyperphoria, with the hope of easing his pain and
improving his sleep.

I distinctly pressed upon the wife the fact that I did *not
think this step would prove in any way curative;* yet I could not
but feel that 4° of manifest hyperphoria was a strain that
ought to be at once removed, especially from so weak an
invalid.

Treatment and Results: As his wife expressed a desire to
try what a correction of his hyperphoria would do for him, a
graduated tenotomy was performed upon the left superior rectus.
The result was a great surprise to his friends as well as myself.
He showed no hyperphoria; esophoria, 0 to 3°; adduction, 25°;
abduction, 8°; sursumduction, right, 4°; left, 4°.

The night following the operation he slept soundly all
night.   He arose the next morning, dressed himself without aid,
and drank three goblets of milk before the rest of the family
were up.   He then sat down and ate a good breakfast, finishing
as quickly as any one.

Within a week he demanded his money from his wife, saying that he would not have her pay his bills for him, and a
short time afterward he began to come daily to the office from
Brooklyn without any one to accompany him.

Two weeks after the operation patient reported that he wrote a long letter (the first in over four months), that he eats well, sleeps well, takes an interest in the newspapers, and is marvelously improved in every way. A full correction of his errors of refraction was ordered and he was instructed to wear his glasses constantly.

Some weeks later a graduated tenotomy was performed on the right internal rectus for the relief of a latent esophoria that had disclosed itself, and the patient returned to Canada to take charge of his business.

For the past five years patient reports that he has had no return of his old symptoms. He has gained over forty pounds in weight and is in better health than for many years. He is again in active charge of his large business interests.

CASE VII.—Mr. O., aged 36; married; clergyman. Referred to me on May 5, 1893.

Family History: Mother had chorea as a child. Two brothers and two uncles suffer severely with headaches.

History of the Case: From a child until 18 years of age the patient suffered with severe sick headaches. From 18 to 32 years of age he was practically free from them.

Four years ago he again began to have severe headaches; which increased in frequency until they became a steady, constant headache. At that time he also began to be *a victim to marked insomnia*, which has been a severe burden for the past two years.

He lies awake for hours before being able to sleep; and often for days at a time gets very little sleep.

He was obliged to give up his ministerial labors for six months, from extreme nervousness bordering closely upon insanity. Any work in his study gave him pain in his eyes and head.

He feared, when I first saw him, that he must again give up his pastorate, as he suffered so severely from the pain in his head and insomnia that he feared insanity.

Eye-defects: Under homatropine patient showed hypermetropia, $+ 1.50^s$. His muscular tests were: Adduction, 20°; abduction, 6°; sursumduction, right, 2°; left, 2°; no hyperphoria; esophoria, $\frac{1}{2}$°. Afterward he developed some latent esophoria.

Treatment and Results: A + 1.00ˢ glass was given for constant wear. In two weeks the patient reported that there had been "a marked improvement in his nervousness and insomnia, but that he had not slept well for the past few nights."

As he showed 1° of esophoria, he was given a 1° prism, and this was increased in strength during the next three weeks as fast as any esophoria in excess of his prism justified such a step. At the end of the three weeks his tests were: Adduction, 29°; abduction, 1° —; esophoria, 6¼°; no hyperphoria.

A graduated tenotomy was now performed upon the right internal rectus and three weeks later his tests were: Adduction, 25°; abduction, 7°; no esophoria; no hyperphoria. He reported that he had had no headache, no feeling of nervousness, and was sleeping well.

A letter dated October 4, 1893 (four months after the operation), says: "I am glad to be able to write you *that I am well.* Just one year ago I offered my resignation as pastor of this church, fearing my days in the ministry were ended. Now I am in excellent health and sometimes almost dare to think I am getting some meat on my bones. Altogether the world looks much brighter since you and my two eyes met." Up to January 1, 1897, no return of the old symptoms had occurred (a period of relief of over three years).

CASE VIII.—Mr. E., aged 59; married; manufacturer. Referred to me on June 22, 1892.

Family History: Father had severe sick headaches. Mother had sick headaches and committed suicide. One maternal uncle was insane, and mental disturbances were common on his mother's side. All his brothers have sick headaches, and one died of phthisis.

History of the Case: Patient had severe sick headaches from boyhood until he was 44 years old. They occurred about once a week.

For the past twelve or fourteen years he has suffered from a distressing gastric trouble, consisting principally of the raising of enormous quantities of gas and often of food after meals.

Accompanying this he has had *extreme* and *chronic insomnia,* often being unable to sleep for an entire night and always sleeping poorly. Great mental depression has followed these troubles.

He has tried during the past ten years almost every form of medicinal and electrical treatment without benefit to his gastric trouble or insomnia.

Eye-defects: At the first visit to my office the patient had $\frac{20}{20}$ vision in each eye and tolerated $+$ 0.50^s. His muscular tests were: Adduction, 27°; abduction, 6° —; sursumduction, right, 2°; left, 2°; no hyperphoria; esophoria, 2°.

Within a week he showed esophoria of 9°, and later a very high degree of latent esophoria; presbyopia, $+$ 3.00^s.

Treatment and Results: The treatment of this patient consisted of three graduated tenotomies, at intervals, during the next three months,—two on the right and one on the left internus.

Ten days after the last operation he showed: Adduction, 60°; abduction, 8°; sursumduction, right, 2°; left, 2°; no hyperphoria; esophoria, $\frac{1}{2}$°.

Soon after the second operation he began to notice a marked improvement in his physical condition. He was in much better spirits. The belching of wind had nearly stopped and he began to sleep fairly well.

December 1, 1893. Patient reports that he is "absolutely well, in spite of heavy business losses and anxieties." These, under ordinary circumstances, would have prostrated him in bed. He states that he sleeps "like a child most of the time, and can digest anything."

Up to June 1, 1896 (four years after the eye-treatment was begun), no return of his old symptoms had occurred. He called at my office to state "that the cure was perfect and apparently permanent, in spite of extreme business reverses and other causes for extreme mental worry and anxiety that he could not have endured but for his physical change."

CASE IX.—Miss B., aged 22. Referred to me on June 6, 1891, by Dr. J. R. Nilsen, of New York.

Family History: No hereditary tendencies to nervous diseases, so far as known.

History of the Case: When 15 years old this patient began to have severe headaches whenever she took any exercise or used her eyes for any length of time.

About four years ago she *began to have extreme insomnia*, and her eyes troubled her so that she was obliged to give up her position as a typewriter. Since then she has not been able to use her eyes, even for a few minutes at a time. She has been skillfully treated for uterine troubles, but no change in her physical condition has been observed.

Two years ago she had nervous prostration, which lasted for six months. At the present time the patient has almost constant headache, which has been somewhat relieved during the past year by $+ 0.75^s$ glasses; she also suffers from very severe attacks of neuralgia every few weeks. Insomnia exists as a habit, so that she never sleeps more than four or five hours, and, usually, does not sleep at all one or two nights each week. There is absolute inability to use her eyes for reading, sewing, or any close work.

Eye-defects: At the first visit patient showed: Adduction, 6°; abduction, 14°; no hyperphoria; exophoria, 9°; crossed diplopia with the red glass. Under atropine she had hypermetropia, $+ 1.25^s$. Later on she developed an extremely high degree of latent exophoria.

Treatment and Results: The hypermetropia was fully corrected by glasses ($+ 1.25^s$), which were given for constant wear. During the next eighteen months four graduated tenotomies were performed on the external recti, and an advancement was performed on each internus.

After the first four months of treatment the patient reported " that she had gained twenty-four pounds in weight; had only had one neuralgic attack; had been sleeping much better, but still had considerable headache."

November 3, 1892. A letter received from the patient says: "I am feeling well, have no headaches, and astonish people by the wonderful improvement in my health. I still have, however, considerable pain in my eyes at times, and am unable to use them for any length of time."

November 10, 1893. Patient visited my office and reported that she has had no neuralgia for six months and very little headache. She sleeps well, except after any unusual excitement or fatigue and weighs twenty-five pounds more than when treatment was begun. She is not able, as yet, to use her eyes more than a half-hour at a time; but even this is a marked improvement over her former asthenopia.

Her eye-tests were: Adduction, 41°; abduction, 8°; no hyperphoria; exophoria, 5°.

She was at this time unable to remain in the city for further eye-treatment that was clearly demanded by some exophoria that still remained uncorrected.

Later reports from this patient indicate that she is practically well, except that she still has some asthenopic symptoms.

CASE X.—Mr. M., aged 48 years; married; merchant. Referred to me on December 29, 1891, by Dr. W. C. Stone, of Lakewood. N. J.

Family history not taken.

History of the Case: This patient had suffered all of his life with headache. These had become very frequent and severe during the past four years.

At the time of his first visit to my office he began to be troubled with confusion of thought, inability to apply himself to business, and *persistent insomnia.*

A few months before he had broken down completely. He was then obliged to give up business and go to Lakewood for rest and treatment.

In spite of the change of air, absolute rest, and medication of various kinds, he did not improve and was unable to sleep.

Any excitement, imprudence in eating, or the smoking of a single cigar was sure to cause a sleepless night.

He was told by his physician, in New York City, that, if he were not benefited by treatment at some sanitarium he must settle up all business matters and devote himself to traveling.

As he did not seem to improve, he was sent to me by a physician at Lakewood, who had been cured of similar symptoms by me, to see whether any eye-defect was the cause of his trouble.

Eye-defects: At the first visit this patient showed: Adduction, 31°; abduction, 4°; no hyperphoria; esophoria, 4°; hypermetropia, + 1.75ˢ (under atropine). Within a week he showed esophoria of 8° and no abduction.

Treatment and Results: Hypermetropic glasses (+ 1.50ˢ) were ordered for constant wear, with a stronger glass for reading. A graduated tenotomy was performed on the left internal rectus about a week later. The improvement in this case was very marked from the first.

At the end of the first week, after wearing his spherical

glasses with esophorial prisms, he reported: "*I have slept well every night.*"

Two weeks after the operation he said that he was steadily improving in strength of body and mind; that he slept splendidly; that he had no confusion in his head; and that he was able to attend to business without any mental or physical fatigue.

March 1, 1892. He has had no return of his old symptoms and is attending to business regularly. His tests are: Adduction, 64°; abduction, 8°; no hyperphoria; no esophoria.

February 27, 1897. This patient states that he has had no return whatever of his old symptoms during the past five years, and that he is actively engaged in his large business enterprises without any physical ills.

CASE XI.—Mrs. M., aged 45; has no children.

Family History: Both parents living and well. Father formerly had some headaches. One brother is insane. Maternal grandfather was insane and committed suicide.

History of the Case: Patient was perfectly well until 20 years of age, when she had her first child. She then began to have headache about three times a year. Six years ago her headaches became very severe and very frequent.

At the present time she averages at least one severe headache every two weeks, during which she is prostrated for two days or more. These headaches are always accompanied by vomiting, which is almost constant for hours. Patient is very nervous and has a constant tremor of the head.

She *suffers severely from insomnia*, sleeping very poorly nine nights out of ten, and seldom closing her eyes before 1 or 2 A.M. She awakes with extreme nervous fear.

Every day she has a feeling of heat, which begins in the chest, extends down to the feet, and then up again. Her spine is very sore to the touch and pains her severely at times. She has always been troubled with constipation.

She has consulted many physicians of eminence during the past five years without any apparent benefit, her insomnia and pain seeming to be beyond the power of drugs.

Eye-defects: When this patient came to me she was wearing + 0.75^s for distance and + 1.50^s for reading, given to her by an oculist in the West whom she had consulted. As she only showed hypermetropia of $+ 1.00^s$ under atropine, these glasses were not altered.

Her muscular tests were: Adduction, 20°; abduction, 8°—; sursumduction, right, 2° +; left, 2°; right hyperphoria, ½°; no esophoria. Within a week she manifested: Right hyperphoria, 4½°; right sursumduction, 6°; left sursumduction, 0. Later still more latent hyperphoria became manifest.

Treatment and Results: A graduated tenotomy was performed on the right superior rectus in one week from the time of her first visit to my office.

She began to improve almost immediately after the operation, and on the third day reported that she felt better than for years, was much less nervous, her appetite was improved, and she had slept thirteen hours the night before.

Two weeks after this she reported *that she had slept eleven or twelve hours every night for the past ten days,* and had had no headaches or morbid fears. The tremor of the head had disappeared entirely.

About three weeks after the first operation the left inferior rectus was "button-holed," the patient in the meantime having worn prisms to correct her hyperphoria as fast as she disclosed it.

Three weeks after the last operation she reported that she "had only had one headache and that was after two days' sewing and general 'overdoing.'"

Her muscular tests at that time were: Adduction, 28°; abduction, 7° +; Sursumduction, right, 3°; left, 3°; no hyperphoria; no esophoria.

July 20, 1893. A letter from the patient states that she has "had no severe headaches, has gained in weight, and is sleeping splendidly."

August 9, 1893. A letter from her physician says: "I am happy to state that your patient is still improving, going up gradually. I never saw a change in a person to equal it. She is the wonder of all those that know her. People meet her on the street and do not recognize her. They say: 'Why Mrs. M., you look fifteen years younger!' I must say I was rather skeptical, but am so no longer."

CHAPTER VI.

EYE-STRAIN AS A CAUSE OF CHRONIC GASTRIC AND DIGESTIVE DISTURBANCES.

No forms of human suffering are harder to bear, or, perhaps, more frequently encountered by the physician, than those associated with the abdomen.

Even when the distress is not severe, but long continued, a marked effect upon the mental and physical state is commonly developed that peculiarly unfits the sufferer to bear his burden with composure or courage.

Furthermore, the intimate connection that exists between the stomach and intestine, on the one hand, and the brain, lungs, and heart, on the other, is well known; hence it is that extreme pallor, faintness, difficulty in breathing, and irregularity of the heart's action often accompany dyspeptic conditions at times and cause the patient proportionate anxiety and alarm.

One patient of mine, whose case is reported in preceding pages, had for many years periodical attacks of intense pain in the stomach accompanied by symptoms of impending death from almost total arrest of the action of the heart and lungs.[1]

I have seen many cases (where the diagnosis of chronic heart disease had been repeatedly made by physicians of repute) which proved to be purely functional disturbances of the heart.

Some years ago one of the most confirmed sufferers from asthma that I ever saw owed his attacks of suffocation entirely to distension of the stomach from gas,—as a result of decomposition of food. His lungs were entirely free from disease.

The percentage of sufferers from chronic digestive disturbances to the total adult population in almost all countries and climes is very large. The stomach and intestine of adults seem particularly apt to become objects of solicitude among the brain-workers. This is comparatively infrequent among those who labor with their muscles alone. This statement applies to the human race with equal force, irrespective of climate, nationality, sex, occupation, or surroundings.

---

[1] This case was one of chorea, accompanied by alarming attacks of gastralgia and sixteen years of nervous prostration.

I mention this fact (whose accuracy cannot, I think, be disputed) because it has often been used by others as the basis of an argument to prove that a *lack of proper exercise* alone entails upon the brain-worker most of his digestive disturbances; and, coupled with this chief cause, sedentary habits, excessive and irregular feeding, alcohol, tobacco, and the large expenditure of nervous vitality in mental processes also add their weight to the eventual collapse of the digestive function.

This argument would seem highly rational to most minds, and is undoubtedly true; yet we find that it cannot be applied to all cases, and that it affords no basis of cure to some sufferers who have fought in vain, by exercise, abstinence, careful living, intelligent medication, and rest, for a cessation of their ills.

It might be argued in rebuttal that the so-called "laboring classes" (where the use of muscles, not brains, brings a revenue) eat most imprudently, hastily, and enormously, as a rule; that they too frequently bolt their food; that they drink large quantities of water with and between their meals; that they are even more apt to be excessive in the use of alcohol and tobacco than the brain-worker; and that they violate in many other ways the commonly-accepted views of how to live. Their tenements are not always sanitary. They often suffer from cold, exposure, and a lack of proper cleanliness.

In spite of these serious drawbacks to health, is it, then, open air and exercise merely that keep, in this class, the stomach, intestine, liver, and pancreas alive to the proper performance of their functions? Do the laboring classes never have digestive troubles? Are flatulency, headaches, eructations from the stomach, obstinate constipation, heart-burn, palpitation of the heart, and nervous prostration never encountered among the muscle-workers? If so, is there no other cause but the imprudence of living and eating to account for these symptoms in all classes? Is it not possible that some factor, not yet properly recognized by most medical men, may contribute (to a large extent) to the downfall of the apparatus of digestion? Cannot it be possible, and is it not even more than probable, that an *exhausted nervous vitality precedes the appearance of digestive ills* in those cases where no actual disease of the stomach, liver, or intestine can be said to exist? May it not be that the digestive organs fail to act properly simply because there is not sufficient nerve-force to run them?

To these questions, which must arise to all thinking men, whether doctors or laymen, many others might be suggested here before an attempt is made at reply.

If it can be shown that *defective nerve-power may cause healthy organs to act imperfectly and spasmodically* in the human body,—just as a low steam-pressure in a mill would affect the running of complicated machinery,—is it not wise to correct that defect before the machinery (or organs) be tampered with ?

If it can and has been scientifically demonstrated that a *leak of nerve-force* often exists in patients whose digestive functions have for years been faulty, is it not best to stop that leak ? We can, then, look for an improved condition of those organs, without drugs, as soon as the reservoir of nerve-force has been filled by nature to a proper level.

I have brought up some of these points as introductory matter to some very interesting cases that I purpose to present. In order to prepare the reader for arguments that I first deem it wise to make, in explanation of my views, I shall, perchance, repeat some things that have already been said by myself in relation to the cure of even more serious nervous disturbances than digestive ills.

But, as some general axioms apply alike to all the varied types of nervous debility and excitability, whether it be epilepsy, chorea, headache, insanity, neuralgia, or functional digestive ills, I may be pardoned if I reiterate certain physiological facts in defence of the following propositions:—

1. EYE-STRAIN [EITHER FROM ERRORS OF REFRACTION UNCORRECTED OR FROM SOME ABNORMALITY OF ADJUSTMENT OF THE EYE-MUSCLES] MAY, AND OFTEN DOES, CAUSE CHRONIC AND INTRACTABLE DISTURBANCES OF THE STOMACH, LIVER, AND INTESTINE.

2. CHRONIC INTESTINAL AND GASTRIC DISTURBANCES MAY BE, AND OFTEN ARE, COMPANIONS TO MUCH GRAVER DISTURBANCES OF THE GENERAL NERVOUS SYSTEM THAT ARE DUE TO THE SAME CONDITION,—VIZ., EYE-STRAIN.

3. EYE-STRAIN MAY EXIST WITHOUT SYMPTOMS REFERABLE TO THE EYES THEMSELVES OF WHICH THE PATIENT IS CONSCIOUS.[1]

---

[1] This important point is fully discussed in my reply to a critical article by Prof. Casey A. Wood. In Professor Wood's article my views were discussed and exception taken to this particular statement.

The reader is referred to the New York Medical Journal, September 29, 1894, in which I give many cases to demonstrate the accuracy of my statement.

Ten years ago the medical profession would have been less prepared than to-day to receive three such startling propositions, even for argument.

To assert that the eyes, and they alone, are sometimes responsible for chronic intestinal and gastric disturbances that have failed to yield to medicinal treatment, massage, rest, travel, restricted diet, and the whole category of recognized methods of treatment may seem, at a first glance, to some of my readers, as incredible.

To advance the view that, in *the brain-workers, as a class, the eyes are a factor worthy of most careful study and patient observation whenever the nervous mechanism seems out of joint* may not have as much weight now with some of my readers as it is sure to have in the future.

The conditions that exist in the two classes of mankind— the brain-workers and the muscle-workers—are radically different from the eye stand-point. In the former the eyes are almost constantly employed within the limit of accommodation ; in the latter the eyes are seldom used in the same way or under as trying conditions.

Does every physician think, when he advises an adult patient "to rest and travel" or a child "to be taken from school," that much of the benefit that may follow is probably due to the relief from eye-work ?

The *expenditure of nerve-force* is much less in using two eyes (that may be defective in focus or badly adjusted for binocular vision) for objects several feet or more distant than when used for reading, sewing, study, drawing, etc. ; much more at the theater, in crowded streets, or where objects are numerous and constantly changing position than in the peaceful country, where the eyes gaze at green fields and distant mountain-tops, or on ship-board, when the horizon is scanned.

One feature in the so-called "rest cure" that is not generally or properly appreciated is the *entire and enforced cessation of all eye-work.* It contributes as largely to the recovery of the patient (in my opinion) as the massage, electricity, and the recumbent posture. These patients are not allowed even to read letters from home ; but they are allowed to be read to by the trained nurse, for mental pastime.

In concluding this chapter the author would draw the

following conclusions respecting the eye-treatment of chronic digestive disturbances :—

1. *Obstinate chronic constipation* is often due to a condition of the eyes that entails a leakage of nerve-force.

Irrespective of the apparent absurdity of this statement to the casual reader (at the first glance), the clinical records of a very large number of patients in the author's case-books (cases where cathartics had been required almost daily for years) positively demonstrate that permanent relief from this distressing and common ailment has followed the correction of eye-strain, and that drugs were no longer required to move the bowel.

The eyes have, in themselves, nothing to do with the digestive organs or the intestinal tract; but an arrest of a leakage of nerve-force can assist materially in furnishing additional power to the organs of digestion. The best machinery will not run properly when steam-pressure is below the required standard, nor will healthy organs in the human body perform their proper functions when the nerve-power is constantly at a low ebb.

2. The percentage of *chronic dyspeptics* that owe their sufferings chiefly to eye-strain is very large, in the opinion of the author.

It certainly can do no harm to this class of sufferers to examine with care the refraction of the eyes and the adjustment of the ocular muscles while they are being treated by drugs, restricted diet, washing of the stomach, etc.

3. Attacks of *intense gastric pain*, even when accompanied by vomiting, eructations of gas, difficulty of breathing, etc., justify an examination of the patient for some form of eye-strain.

Two of the worst cases of this type of suffering that the author has ever seen are reported in full (in pages 122 and 167).

4. *Vomiting*, whether preceded or accompanied by headache or when independent of pain, points strongly, in the male, to the existence of some reflex eye-cause, and often in the female. This suggestion applies particularly to those who suffer from " weak stomach " and a tendency toward nausea. In females the possibility of pregnancy and of reflex pelvic disturbances should also be considered.

CASE I. *Chronic Dyspepsia of Fourteen Years' Duration.* —Mr. E., aged 61 years; manufacturer; married; has two children.

Family History : Father had frequent sick headaches. Mother had sick headaches and was thought to be insane at times. Mental disturbances were very common among his maternal ancestry. One near relative was deranged for years. All of his brothers have sick headaches and one died of phthisis. One son is a dipsomaniac.

History of the Case : This patient suffered severely from sick headaches (on an average of one each week) from boyhood until he was over 40 years of age.

For the past fourteen years he has suffered terribly from gastric disturbances, especially from raising of gases and food after each meal.

For this trouble he has tried almost every form of medical and electrical treatment without relief. His sufferings have always been laid to a " torpid liver " by most of the physicians consulted by him.

He suffers also with extreme mental depression and has had fears of hereditary insanity.

Eye-defects : On the first examination this patient had $\frac{20}{20}$ vision in each eye, but tolerated $+ 0.50^s$. He required $+ 2.75^s$ for reading. His muscular tests were: No hyperphoria; esophoria, 3°; adduction, 29°; abduction, 6° +; sursumduction, right, 2°; left, 2°. Subsequently a very high degree of latent esophoria was discovered.

Treatment and Results : The patient wore prisms during the summer, with decided improvement in his digestion and a marked buoyancy of spirits that contrasted markedly with his previous despondency.

A graduated tenotomy was then performed upon each internal rectus. A very rapid improvement in his stomachic disturbances immediately followed.

One month later the right internus was again operated upon. The patient continued to steadily improve in physical condition, mental condition, and digestion.

The last report, received recently, stated that he was perfectly well, slept well, was in good spirits, and *had not had the slightest trouble from his stomach for nearly four years.*

His eye-tests to-day are : No hyperphoria ; esophoria, ½° ; adduction, 50° ; abduction, 7°.

CASE II. *Chronic Indigestion, with Fainting Attacks.*— Mr. E., aged 35 ; merchant ; married.

Family history not known.

History of the Case: This patient had been subject all of his life to a " weak stomach " and a tendency to faint very easily.

The symptoms of chronic indigestion increased in severity. During the past year they were so aggravated that he lost thirty pounds in weight.

He has suffered also with asthenopia at times.

Eye-defects: At the first visit, September 20, 1893, the patient had the following tests: O. D. V., $\frac{20}{15}$ — (tolerates no + glass); O. S. V., $\frac{20}{40}$, made $\frac{20}{15}$ — by + 0.75° (axis. 90°); adduction, 21° ; abduction, 7° ; sursumduction, right, 3° ; left, 3° ; no hyperphoria ; no esophoria. Under homatropine he showed : O. D. V., $\frac{20}{40}$, made $\frac{20}{15}$ — by + 0.50 $\bigcirc$ + 0.50° (axis. 90°); O. S. V., $\frac{20}{70}$, made $\frac{20}{15}$ — by + 0.50 $\bigcirc$ + 1.00° (axis, 90°).

Treatment and Results: The following glasses were ordered for constant wear. September 20, 1893: O. D., + 0.50° (axis, 90°); O. S., + 1.00° (axis, 90°). Later a full refractive correction was ordered for constant use.

On November 15, 1893, the patient reported that he had *gained ten pounds in weight,* had no gastric disturbance, and had felt very well. Eye-tests: O. D. V., $\frac{20}{15}$; O. S. V., $\frac{20}{15}$; adduction, 60° ; abduction, 8° ; no heterophoria.

July 10, 1896. Patient reports that he has " *gained forty pounds in weight,* has no gastric disturbance, and is perfectly well." All symptoms of his old fainting attacks have entirely disappeared.

CASE III. *Periodical Attacks of Uncontrollable Vomiting, followed by a Total Loss of Power and Sensation in the Arms.* —Miss M., aged 24.

Family History: Parents not overstrong. Some hereditary predisposition to weak digestion.

History of the Case: Ten years ago the patient had a sudden attack of violent pain in the head, followed by vomiting for several hours. She lost co-ordination of the muscles of the tongue; had numbness of the hands and arms; and, finally, lost all power in her arms, which never fully regained their normal strength until the eye-muscles were treated. For ten years she has had (about once in two weeks) an attack beginning with numbness in some part of the body (usually the arms) followed by pain in the head and persistent vomiting for hours.

Two years before I saw her she had her eyes examined, and was given for constant wear spherocylinders to correct her hypermetropia and astigmatism.

Since putting on her glasses she has had about one of the attacks described each month, much less severe and without pain in the head. Each attack begins with numbness, and this is followed by vomiting, blurring of vision, pain in the eyes, intolerance of light, and great difficulty in breathing.

Two years ago she was told by her physician that she had "hasty consumption," and a very unfavorable prognosis was given. She went to Lake George for two seasons, and this tendency seems to have disappeared under climatic treatment alone.

Eye-defects: At the first visit to my office (May 6, 1892) the patient was wearing constantly the following glass: O. D., $+ 1.75 \, \bigcirc + 0.75^c$ (axis, 90°); O. S., $+ 2.00 \, \bigcirc + 0.50^c$ (axis, 90°). With these glasses on her tests were: O. D. V., $\frac{20}{15} -$; O. S. V., $\frac{20}{15} -$; adduction, 29°; abduction, 5°; sursumduction, right, 2°; left, 2°; no hyperphoria; esophoria, 2°. Under atropine the glasses were found to be a full correction for her hypermetropia and astigmatism, and no change was made in them. A *very high degree of latent esophoria* was subsequently disclosed.

Treatment and Results: Prisms were put on at once for esophoria, and, during the next two months, three graduated tenotomies were performed on the internal recti for a very high degree of latent esophoria. She had no attack during this period,—a longer interval than had occurred for ten years.

During the four years that have elapsed since the eye-treatment was completed this patient has gained nearly twenty pounds in weight, and has been almost entirely free from headache or vomiting.

She has *entirely regained the power of her arms, and has no longer any loss of sensation or numbness in her extremities.* Her recovery seems to those that knew her when ill as almost miraculous. She can use her arms for sewing. embroidery, etc., as well as if she had never had any loss of power.

CASE IV. *Chronic Indigestion. Vomiting, and Nervous Prostration.*—Mr. C., aged 25 ; single ; divinity student.

Family History : Mother has headaches and has had pulmonary hæmorrhages. Father has neuralgia of the stomach. Maternal grandmother and two uncles died of phthisis.

History of the Case : This patient has always been delicate. He has had headaches and periodical attacks of vomiting and indigestion since childhood. He ran down rapidly during his senior year in college with gastric disturbances and headache.

When he came to me (February 28. 1888) he was suffering severely from impaired digestive functions, irregular action of the bowels, alternating constipation and diarrhœa, headache slow heart's action. cold extremities, and very marked physical weakness. Everything seemed to point toward an attack of complete nervous prostration.

Eye-defects : Under atropine this patient's refraction was : O. D., $+ 0.50^s$; O. S., $+ 0.75^c$ (axis. 112°). His muscular tests were : Adduction, 37° ; abduction. 6° ; sursumduction, right, 3° ; left. 3° ; no hyperphoria ; esophoria. 2°. Subsequently about 20° of latent esophoria were disclosed.

Treatment and Results : The treatment in this case consisted of a graduated tenotomy upon each internal rectus and the use of a full refractive correction for reading.

The improvement in this case began almost immediately after the first operation, and five months later the patient reported that he had gained over fifteen pounds in weight, had no dyspepsia. had regained his physical strength almost completely, and was in better health than for many years. He still had some asthenopic symptoms, if he used his eyes at night.

December 14, 1888. The patient reports that he considers himself cured. He has no dyspepsia, has gained twenty pounds in weight, and can use his eyes with perfect comfort.

He has remained well for the past eight years. His tests to-day are : Adduction, 64° ; abduction. 7° ; sursumduction, right. 3° ; left, 3° ; no hyperphoria ; no esophoria.

CASE V. *Chronic Constipation. Intestinal Hæmorrhages, and a Complete Nervous Collapse.*—Mr. H., aged 46 ; manufacturer ; married.

Family History: Both parents lived to old age. Two paternal uncles died of phthisis. No hereditary tendency to nervous diseases seems to exist.

Eye-defects: Vision, $\frac{20}{20}$, without atropine. Under atropine, a latent hypermetropia of $+ 1.00^s$ in each eye. Patient has never required a glass for reading. Manifest esophoria, 5°. After using prismatic glasses for a short time the patient showed: Esophoria, 13°; adduction, 24°; abduction, 4° —. Later on the adduction exceeded 50° and the abduction fell to 0. Homonymous diplopia with the red glass over one eye was usually present; and at times, without the red glass, it annoyed the patient.

History of the Case: This patient had been a perfectly well man and had carried on a very large business up to fifteen years ago.

For many years he has been unable to look out of a car-window while traveling without great distress.

He had suffered all his life from obstinate constipation, and had taken cathartics so regularly that any cathartic water caused intestinal hæmorrhage prior to the eye-treatment.

When this patient first came to me he was able, by the most careful diet, regular habits, and by retiring at 8 or 9 o'clock, to carry on his enormous business only with the greatest difficulty because of the following symptoms: Extreme nervousness after the slightest fatigue and marked mental depression without any cause.

Treatment and Results: The treatment of this patient consisted, at first, of the wearing of prisms to relieve the esophoria, and, later on, of graduated tenotomies on both internal recti. Subsequently $+ 0.50^s$ glasses were given for constant wear, and $+ 1.00^s$ glasses for near work. The improvement in his condition was marked and continuous from the first, and he writes that he is so busy and feeling so well that he cannot find time to have the slight remaining esophoria corrected. Over five years have now elapsed without any return of his former ill health, during which time he has constantly been engaged in active business pursuits.

His chronic constipation and tendency toward intestinal hæmorrhage have totally disappeared. He has gained about forty pounds in weight.

CASE VI.—Mr. W., aged 55; married; manufacturer.

Family History : One daughter has for years been suffering from nervous prostration, with extreme depression bordering upon melancholia. Dyspeptic troubles very common in ancestry; also severe " bilious attacks."

History of the Case: This patient had suffered most of his life from chronic dyspepsia and bilious attacks that had failed to be benefited by any form of medication. He had tried travel, rest, restricted diet of all kinds, electricity, massage, etc.. without appreciable benefit.

Eye-defects: At the first visit, on December 7, 1894. the eye-tests were : O. D. V.. $\frac{20}{15}$ —, slightly improved by $+$ 0.50^c (axis, 165°); O. S. V., $\frac{20}{30}$ —, made $\frac{20}{15}$ — by $+$ 0.50^c (axis. 165°); adduction, 19°; abduction. 5°; no hyperphoria; esophoria. $\frac{3}{4}$°. By the use of prisms about 12° of latent esophoria were disclosed.

Treatment and Results: The treatment in this case consisted of one graduated tenotomy on each internal rectus.

A letter from his daughter, dated January 13. 1896. says: " Father has been wonderfully well for some months. the past summer being the most comfortable one he has spent for years."

CHAPTER VII.

THE EYE-TREATMENT OF EPILEPTICS.

No sadder misfortune can befall a human being than to develop epilepsy. The horrible aspects of this disease are not confined alone to those who are called upon to witness the attacks or to administer to the apparent sufferings of the afflicted. Their anxiety and distress are naturally most acute for the few moments during which they see a fellow-being suddenly become unconscious of his surroundings; distorted in face, trunk, and limbs; frothing at the mouth, livid in the countenance, and often bleeding from some wound produced by the sudden fall that ushers in the attack. But, to the sufferer from this dreadful malady, the distress of mind is not momentary, nor do the alarm and unconquerable dread disappear when the re-establishment of consciousness and physical power follows an attack.

Sooner or later an epileptic must become aware that his attacks are horrible to witness, that they are liable to occur at very unfortunate and unexpected moments, and that it is politic to keep people in ignorance of the existing malady.[1]

For awhile such victims are buoyed with a delusive hope that drugs will eventually restore them to health, and for a period varying from a few months to many years they are faithful devotees to some special combination of the bromide salts that their medical adviser or advisers may select. As each attack occurs, or as different physicians are consulted, some slight modifications of the bromide formulæ are commonly made; once again the patient's waning courage and increasing despair are bolstered by words of encouragement and the oft-repeated assurance that " this special combination will probably act better than the last one "; again, the patient goes on, too often conscious of a decreasing mental power, a failing memory, or a general bromide stupor, and usually with a disfigured face from bromide eruptions that advertises his sad malady to the

---

[1] To relieve these subjects from enforced social ostracism and to improve their mental condition, the establishment of epileptic colonies has been suggested and a few are now in operation.

world; again and again his slavedom to drugs is re-established
with professional concurrence, when everything in the clinical
history of the patient cries aloud to stop; and during all this
time—the years of drugging, mental deterioration, and despair
—no attempt has, in most cases, been made or even suggested
that sources of reflex irritation of the nerve-centers be sought
for by methods of scientific precision, and, as far as possible
relieved before the constitution of the patient is broken down
by drugs.

This is not an overdrawn picture nor an exaggeration of the
facts, as I see them almost daily in my consultation-rooms.

That certain patients can and do take moderate doses of
bromides (in all possible combinations) for a time without any
very marked injury to health, and often with marked controlling
influences upon epileptic seizures it is useless to deny. Were
it not so, the general use of these drugs could not have become
so firmly established in medical therapeutics. The bromide
salts (when first introduced to the medical profession) so far
exceeded, in their results upon epileptic seizures, the preparations of zinc, silver, borax, etc., then in vogue, that the enthusiasm of the profession over a possible specific for epilepsy has
been dissipated more slowly than it otherwise would have been.
It has taken many years to convince most of the leaders of professional thought in this country and in Europe that too much
should not be expected from the bromide salts; that the beneficial results obtained from them in epilepsy are not permanent,
as demonstrated by the frequent attacks that follow a complete
cessation of the drug; that the constitutional effects of bromide
salts may, in chronic cases of epilepsy, prove serious; and that
epilepsy is an incurable disease when treated exclusively by
drugs, except in the rarest of instances.

The point that I would make in this connection regarding
the administration of the bromide salts is that the *principle of
treatment of convulsive seizures by drugs is wrong*, provided no
previous attempts have been made to discover or remove conditions that can be accurately determined and which have been
shown, by the observations of more than one, to be related to
the causation of epilepsy.

I can see no more intelligence in administering bromides to
an epileptic patient until the condition of the eyes—and often
of the nose, teeth, genitals, and anus—has been carefully ex-

amined for positive sources of reflex nervous disturbance, than I should in administering chloroform regularly to a patient afflicted with a severe toothache to stop the manifestations of an abscess at its root or some exposed cavity.

To *benumb the cerebral centers by a drug* so that they cease to respond to reflex irritation from any peripheral source *is not curing the disease.*

Such a principle of treatment is absurdly irrational and opposed to physiological research. Experiment has clearly demonstrated that epileptic seizures may follow peripheral lesions from the reflex irritation thus excited. This cannot to-day be denied in the face of overwhelming evidence. To treat epilepsy by drugs without previously searching for all the exciting causes that may exist is no longer creditable to the medical attendant.[1]

On the other hand, while discussing any form of treatment of epilepsy, we are, as a rule, compelled to accept the results of bromide salts upon the number of seizures in any individual case as a basis of comparison, when the results of any other method are subsequently recorded.

Bromide salts (in various combinations) are to-day generally regarded by the medical profession as certainly the most prompt, and possibly, for a time, the most efficacious form of treatment of epilepsy.

When it can be shown that in any individual case the number of epileptic seizures has been made numerically less than when bromides have been skillfully employed for a sufficient period to test their efficacy, or, in other words, if it can be demonstrated that the condition of any epileptic patient has been improved without the aid of drugs, no medical man can deny that an important advance has been made in the scientific treatment of this malady.

Those who have had the largest experience in the use of bromides are to-day beginning to lose faith in them as a curative measure in the treatment of epilepsy.

Within the past year I have seen many patients with epilepsy who have been advised by leading medical men of the

---

[1] The reader is referred to my reply to Dr. Frederick Peterson (New York Medical Journal, December, 1896, and January, 1897) for many *important pieces of documentary evidence* relating to the cases reported in connection with this chapter. Letters from patients and their physicians are given which fully substantiate the accuracy of my reports of these cases. Many other controversial paragraphs can be found in that reply which cannot be reproduced in a book.

larger cities to consult an oculist, and it is the exception, so far as I can learn, for medical men of distinction to-day to extend to an epileptic patient any hope of permanent cure by medication alone.

Not long since a leading neurologist of this country told a patient afflicted with epilepsy that "bromides, trephining, and eye-treatment comprised about all that offered any chance of his improvement."

It is well known that a total cessation of bromides in many instances where they have been administered for any length of time as an aid in controlling epileptic seizures is followed by a very marked increase in the frequency and severity of the attacks. So marked is the relapse in some cases that the patient, the family, the friends, and even the medical attendant become seriously alarmed, and the drug is again resorted to, in spite of the reasons that led to its cessation, simply as a choice between two evils,—mental and physical deterioration, on the one hand, and *status epilepticus* or possible death, on the other.

One case of this kind comes vividly to my mind in this connection. A young girl afflicted with epilepsy had taken bromide salts for some years with the effect of controlling her attacks to intervals of from ten to fourteen days. She was brought to me for an examination of her case, and very marked defects of the eyes, both of focus and muscular adjustment, were discovered. I advised the friends to discontinue the bromides for a month, during which time I might be able to form some opinion respecting the frequency of her attacks without drugs and have a fair basis for comparison regarding the effects of eye-treatment upon her at the expiration of that period. As I proposed to use no drugs during the correction of the eye-defect, the condition of the patient could then be contrasted from time to time with her actual condition before the eye-treatment, when it was not masked by the influence of drugs.

The parents consented. Within forty-eight hours the child had *continuous convulsions for twelve hours*, and the medical attendant wrote me: "I was so afraid that she would die that I immediately began the administration of bromides in large doses." He asked me in this case to make an exception to my rule and to allow the patient to take bromides until I deemed it safe to discontinue them. After two weeks of eye-treatment all drugs were discontinued without any serious results to health.

A similar result to this was mentioned in a medical meeting some years ago by a leading neurologist of this city as occurring in one of his patients whom I advised to discontinue bromides prior to a proposed tenotomy by me upon one of the eye-muscles. A condition of continuous epilepsy followed within twenty-four hours,—although, while under bromides, the patient had been led to think that a cure was being effected by their use.

In my experience, the cessation of bromides is sure to be followed, in about one-half of the victims to epilepsy, by a very marked increase in the number of seizures, just as a cessation from drinking in a confirmed inebriate discloses to him his actual physical condition. Whether it is wise to continue a habit that is slowly undermining the health, because the victim is afraid to stop, is an important question. Unfortunately it cannot be discussed here. It is imperative, however, in all honest endeavors to determine the results of any scientific procedure, that the results obtained shall not be open to question and invidious criticism from neglect on the part of the investigator to exclude other factors that may have aided in attaining the published results. In all the cases that I report in this chapter no drugs were employed while the patient was under my care, unless otherwise stated.

In looking over the records of those cases of epilepsy that have been subjected to the eye-treatment exclusively while under my care, I find it impossible to incorporate them all. The number, in the first place, would be too large to admit of their complete histories, and the labor of preparing their records for publication would be too great. Again, many of the patients, for personal reasons, refused to submit to such steps as I deemed wise for their epileptic condition and were dismissed by me as patients, while others had organic diseases that precluded them from being fit subjects for the eye-treatment. A certain number had marked evidence of kidney disease, syphilis, and tuberculosis.

Some began treatment, and, after satisfactory progress for a few months, failed to notify me of their change of residence and have not been seen or heard from for a long time. Some have died of intercurrent diseases. A few have been killed by accidents, while ocular anomalies that had not been rectified prior to the accident were being investigated.

While it would be of great scientific interest to obtain some deductions regarding the percentages of progress in those patients that have been under this treatment, I am satisfied that a *purely numerical basis for estimating results in the treatment of epilepsy is impracticable and misleading.*

A patient who (while under bromides) was stupid, apathetic, and incapable of participation in business, has been greatly benefited (even if his attacks are not markedly less than when under bromides), provided he has regained his normal intellectual status and is capable of self-support.

Numerically, he might be regarded as unimproved, if the actual number of his attacks should be counted; but the gratitude of the patient and his friends for his restoration to usefulness is not without reason.

A person who has been enabled to dispense with a constant attendant or nurse has been benefited; a child who can go to school and make satisfactory progress in his classes after escaping the thraldom of bromides is vastly better than a sleepy, sluggish child who has to be constantly cared for, even if he does have an occasional epileptic seizure.

Some years ago, in reply to a somewhat severe criticism upon the views relating to the treatment of epilepsy that were published in my work on nervous diseases,[1] I contributed an article to the *Boston Medical and Surgical Journal* from which I may pertinently quote a few sentences. I remark in this connection : " Josh Billings, in one of his characteristic effusions, says : *' I hav alwus notised when an individual hain't got the ability tew criticise judiciously, he dams indiskriminately.'*

" Relative to the question of treatment of some of the functional neuroses (such as epilepsy, chorea, neuralgia, insanity, hysteria, headaches, etc.) two distinct and opposed views are now held. Those who still pin their faith upon the internal administration of the various bromide salts in these diseases remain apparently indifferent to the fact (which can hardly be disputed in the face of overwhelming evidence) that serious and permanent harm almost inevitably follows their administration in large doses for any length of time. Those, on the other

---

[1] Lectures on Nervous Diseases, The F. A. Davis Co., Philadelphia, 1890.

hand, who argue that scientific measures of precision (when properly employed) will often reveal the existence of an *exciting cause* for these nervous conditions that drugs only tend to alleviate for a time, and who base this treatment upon the accurate determination of the exciting cause and its complete correction before they resort to drugs, are to-day subjected not infrequently to ridicule and personal attack by others not thoroughly posted in the methods employed by them."

Before I pass to the consideration of individual cases of epilepsy whose histories have been prepared by me with some care (not alone with the object of showing the ultimate results of treatment of the eyes, but also of demonstrating that permanent and beneficial effects have been secured in correcting anomalies of adjustment of the eyes by graduated tenotomies upon the eye-muscles), it may be well for me to impress upon the reader an important fact, which seems peculiarly hard for some oculists to grasp,—viz., that *latent anomalies of the eye-muscles are to-day as well proved as is the existence of latent hypermetropia.*

By this I mean that the amount of esophoria, exophoria, or hyperphoria discovered at the first examination of any patient does not necessarily represent the *actual amount that exists.* In many cases it constitutes but a very small proportion of the amount that has to be corrected before beneficial results are to be expected from eye-treatment (see Cases IV, VI, VII, XII, XIII, and others).

I recall an epileptic who manifested, in many visits at my office, but one degree of esophoria; yet that patient eventually disclosed, after weeks of careful observation, a high degree of esophoria, and at times had unconquerable double images (homonymous diplopia).

My friend, Dr. G. T. Stevens, has very happily illustrated this very important point in a paper upon heterophoria,[1] read by him before the Pan-American Medical Congress. I take the liberty of quoting a few paragraphs that bear vitally upon the investigation and solution of complex ocular problems. He says :—

"One of the first of the elements in the management of these cases is patience. The oculist who imagines that he can measure and correct his cases of heterophoria in an off-hand manner has not reached an understanding of the first principles of the subject. A case which at first appears quite simple and easy to handle may prove to be one which, for its best correction, may demand the exercise of great skill and of unwearied patience during many months.

" We cannot know, when we begin the treatment of a case of heterophoria, the full meaning of the indications which we then discover. One who thinks that he has found an instrument which will reveal latent heterophoria is surely laboring under a misapprehension. As we go on in the treatment of these cases new difficulties may arise and new efforts may be demanded. This is no reason for not beginning the work, and less for abandoning it before completion.

" I have sometimes illustrated these progressive difficulties by recalling the experience of one who climbs a mountain. He sees before him a height which appears to him the summit which is to reward him for his effort, but when he reaches it he sees before him another even higher than the one which he has surmounted. He ascends higher, and when, at length, he attains to the second height the real summit is still above him. Is he, therefore, to say: 'There is no end to this, and one might climb indefinitely without finding the summit'?

" There is a summit, and one may reach it if he has the needful strength and courage.

" I am sometimes asked by my colleagues if there is no end to the treatment of these cases. Does not the condition which has been apparently removed return? Must the treatment not continue indefinitely?

" The heterophoric condition which has been, in any measure, properly corrected does not return, but that part of the original defect which was at one time latent may at a subsequent time become manifest.

" There is as surely a limit to this work as there is a final summit to the mountain, and when one reaches that limit he has an infinitely greater satisfaction than he who finally gazes upon the stretch of hills and plains and lakes from the elevation which he has earned.

" Many who do not care to take the trouble to climb decline to accept the reports of those who do, and many an eminent man who has lingered about the foot-hills of this work is incredulous because he cannot obtain the same grand view that has been described by some one who has gone beyond.

"An essential to success in this work must be a thorough knowledge of the principles governing the tensions of the muscles with which we work.

"Instruments cannot know these principles. They are only to be learned by long and patient research on the part of the oculist."

Again, I would endeavor to disabuse the professional mind of an impression that seems to be too common, in spite of the amount of literature that has appeared during the past decade upon this field. *The development of latent muscular anomalies of the orbit is not to be confounded with the effects produced by forcing the ocular muscles to adjust themselves to prismatic glasses that are not properly prescribed.*

There is a radical difference between "forcing a patient to adjust his eyes to a prismatic glass" that he has had given him and assisting a patient in his ocular perplexities by giving him a prismatic glass *that only partially corrects* the amount of heterophoria that he discloses.

So long as any patient accepts a prism for the relief of an existing eye-strain that he discloses, and still shows (while wearing it) a clear surplus that the prism fails to correct, we are simply following the guide that the patient unconsciously gives us as a step toward the solution of the problem without any possibility of "forcing the patient to adjust for the prism." (See page 55.)

Again, if we find (by measuring the strength of the individual eye-muscles) that the amount of deficient power in any one set of muscles corresponds to, and justifies the amount of. abnormal deviating tendency of the eyes that the patient discloses, we are still further strengthened in our position relative to the prismatic glass prescribed.

Case VI illustrates this point as well as any case can; and the reader is referred to the history of this patient and the details of his treatment. This boy had almost *absolutely perfect refraction*, even under atropine; so that all questions regarding his "apparent muscular anomalies being caused by some error of refraction" is refuted in this case.

Moreover, his *heterophoria was extreme, but almost totally latent.* The deviating tendencies of the eyes were both vertical and lateral,—hyperesophoria.

This boy's history has been published by me before as one of the most remarkable cases on record. I am glad to report that, during seven years from the date of his first visit to my office, he has had but seven days during which seizures have occurred, in spite of the complete withdrawal of all drugs, and that he *has passed two years and three months without a seizure.*

Now, what has been done for this boy, thus far, by eye-treatment? He has already passed over eight years without recourse to poisonous drugs; he has been saved thus far from a life in a lunatic asylum and restored to usefulness and health; he has had, except on six occasions, complete immunity from his horrible disease, in spite of the total cessation of bromides; he has returned to his former association with his companions, and he is to-day able to go about without an attendant or the dread of impending disaster and possible confinement.

He has happily learned, I trust, that excessive and violent exercise and a disregard of instructions about wearing his prismatic glasses when ordered by me are dangerous to his comfort, as they tend to cause an epileptic attack and also to derange his digestive apparatus seriously. Had it not been for such extreme imprudence he would probably have been entirely free from attacks.

Do we know that this remarkable change is due to the eye-treatment? Most certainly!

This patient had never before passed so long a time without attacks as he has since eye-treatment was begun, although he has been constantly drugged according to the most approved fashion of the present day for his epileptic seizures. During the year preceding eye-treatment he had had thirty-four days in which a series of fits occurred, although heavily drugged with chloral and bromides. He had found in the bromides for some years the only refuge that medicine offered to keep these frightful attacks within bounds that did not seriously endanger life. He naturally felt, as did his parents, that to let go that anchor was to drift beyond aid into hopeless despair. When, therefore, I stopped his bromides at the first visit, every one concerned (the patient, his parents, his friends, and myself) felt quite sure that, unless something was done as a substitute for the drugs, the fits would surely become more frequent and severe. This substitute for drugs *took the form of an operation* for what I deemed the exciting cause of the attacks. Another operation was done on the corresponding muscle of the other eye as soon as the necessity for it became apparent. Then we felt comparatively safe, and the patient could wait with greater safety the effects of prismatic glasses.

When a house has been partially burned no one expects that putting out the fire will at once restore the house. It does, however, prevent further damage and decreases the time and cost demanded for its restoration. Hence it is always deemed imperative to extinguish the fire without unnecessary delay.

We must all admit, I think, that epilepsy is certainly the gravest of all the functional nervous maladies, and that it is, as a rule, incurable by drugs; hence, as I have remarked in a previous discussion concerning this subject, "one radical cure of epilepsy without the aid of drugs offsets a thousand failures as a scientific proof of a discovery."

Case I again illustrates the fact that a *latent esophoria* (independent of any marked error of refraction) was clearly a factor in producing from *two to ten attacks of petit mal during each day*, which bromides failed to control. He has passed nearly five years without any symptom of epilepsy,—since the date when the esophoria was radically relieved by graduated tenotomies.

Case II shows clearly that *hyperesophoria* was a more important factor in causing epilepsy than astigmatism. No marked relief followed the correction of the latter by glasses, but a remarkable change for the better followed graduated tenotomies upon the interni and the superior rectus muscle.

Case III illustrates a very *remarkable improvement in an epileptic idiot* that followed the correction of double convergent strabismus by two tenotomies and a high degree of latent hypermetropia by glasses. The ignorance of his parents allowed this boy to discard his glasses later on. As a result, he relapsed into his epileptic tendencies.

Case IV illustrates a *complete cure of chronic epilepsy* (that had withstood all therapeutic treatment for twenty-four years) through the correction of latent esophoria by graduated tenotomies and latent hypermetropia by glasses. For ten years, without drugs, no epileptic seizures have occurred.

Case V shows the existence of an *extreme degree of astigmatism in one eye*, associated with a large amount of latent esophoria. The patient has regained his full mental power (that became seriously impaired by the use of bromides), and has only had three slight epileptic seizures in six years following graduated tenotomies upon both interni and a full correction of the astigmatism by glasses. Each seizure was directly attributable to causes that might excite epilepsy even in health.

Case VI has already been discussed at some length. Only one slight seizure has occurred in nearly five years. as contrasted with thirty-four days of almost continuous convulsions during the year preceding the eye-treatment, while the patient was heavily drugged with bromides and chloral.

Case VII illustrates the fact that an epileptic patient. whose family history was not of a promising kind and who had inherited a marked tendency toward nervous disease. was cured through the relief of latent esophoria by graduated tenotomies, and the correction of a moderate degree of latent hypermetropia by the use of glasses for reading and other eye-work at close points. No epileptic seizure has occurred for five years and six months. The direct relationship between the eyes of this patient and his epileptic seizures was shown (somewhat unusually) by a very marked tendency toward an epileptic attack whenever an attempt to measure the strength of the interni was made prior to the dates of the graduated tenotomies for the relief of the esophoria.

Case VIII illustrates simply that *admirable results (as regards adjustment) from graduated tenotomies can be obtained in cases of extreme heterophoria.* This patient, unfortunately, lacked sufficient patience and fidelity to enable me to determine the ultimate effects of the eye-treatment upon his epileptic seizures.

Case IX shows that a sudden withdrawal of bromides from the patient prior to the correction of the heterophoria caused (within twenty-four hours) a state of continuous epilepsy that imperiled life and that, *after two graduated tenotomies, the bromides were suddenly withdrawn from the same patient without any ill effects.* She has fewer attacks to-day than when under the influence of bromides, and is markedly improved in her general physical condition. Photographs of this patient in my possession, made at intervals from untouched negatives prior to and during the eye-treatment, illustrate the physical change in a remarkable way.

Case X proved to be one of *extreme latent* esophoria. Although the power of abduction has twice been apparently brought to nearly the normal point by graduated tenotomies upon the interni, a gradual diminution of the power of the externi has occurred. The patient has been practically cured of his epilepsy, however, by the eye-treatment. For over five years he has been free from attacks of epilepsy.

Case XI was one of severe and frequent epileptic seizures that, unfortunately, was terminated by the accidental death of the patient before the final results of eye-treatment were determined. Sufficient improvement had, however, been observed to justify the belief that still better results would in time have been recorded. The family history in this case was a bad one, and the patient was a difficult one to control, being headstrong and disobedient to instructions.

Case XII illustrates a remarkable restoration to health (after a complete withdrawal of bromides that manifested their poisonous effects to a marked degree) from the correction of eye-defects that were indirectly creating reflex nervous disturbances. Not only have all epileptic seizures ceased for the past four years, but all hysterical attacks, that were formerly so frequent, have apparently been entirely cured.

A letter written by her husband from the West, says: " I am glad to say that Mrs. G. passed through the trip safely, with no trouble of any kind. She did not even have a headache nor feel in the least nauseated, which is something unheard of for her before. She appears bright and cheerful, and we feel that we owe it to you."

Case XIII is one of the most interesting cases, from a scientific stand-point, that I have yet encountered. Under full effects of atropine she showed *no errors of refraction*, and for several successive days *no apparent error of adjustment of the eyes was manifested*. The case looked unpromising, and the cause of her epilepsy seemed to be involved in obscurity. For a week the effects of esophorial prisms upon the patient were observed, a low abducting power being the only guide to the existing latent heterophoria that eventually was disclosed by the patient. A radical correction of the heterophoria (that was totally latent) by one graduated tenotomy has for over three years arrested any tendency to convulsive seizures, and apparently effected a complete restoration to health in many other ways.

Case XIV is not as yet of much value in determining the results of eye-treatment upon epilepsy; but it resembles Case XIII very closely and illustrates that a perfectly normal refraction (emmetropia) may co-exist with a *marked heterophoria almost totally latent.* The terrible accident that befell this patient has delayed until recently further investigation of the heterophoria.

Case XV was an extremely interesting one to me for the following reasons: The epilepsy had been of the nocturnal type; it had not been benefited by bromides; the patient had inherited (as shown by my case-records) marked tendencies. on both the maternal and paternal side, to eye-defects; one sister had been a source of great anxiety to the parents for many years from extreme nervous debility that seemed incurable, and she had been restored to perfect health while under my care through eye-treatment without drugs; finally, one brother had manifested epileptic tendencies. He is now practically cured of his epilepsy. A perfect recovery from a life-long dependence upon cathartics is also another evidence of the improved nervous tone of this patient.

Case XVI is yet under observation. It is one of the most difficult cases to solve that I have ever encountered. Thus far the results of eye-treatment seem not to have been satisfactory.

Case XVII illustrates some excellent results in correcting heterophoria, with apparently negative results upon the epileptic seizures. The seizures are so infrequent that it is difficult to tell just what the ultimate effects of the eye-treatment may be. I am inclined to suspect that some form of uterine or ovarian reflex irritation exists at present in this case.

Case XVIII shows that excellent results from the oculistic stand-point have been obtained, with equally marked benefits to the patient in respect to her epileptic seizures.

Case XIX illustrates extreme and unequal refractive errors in addition to extreme heterophoria. The results have thus far been quite encouraging, but the patient has not given me an opportunity to complete my work.

Case XX shows that results have followed eye-treatment which have never been obtained by drugs. This patient has had no epileptic seizures for nearly four years.

Case XXI is not as much improved as some of the preceding cases, but he is as well without drugs as he ever was when under their influence.

Part of the operative work in this case was done before I saw the patient; hence I know but little of what the heterophoria originally was.

Cases XXII and XXIII again illustrate one important point,—viz., that *very extreme latent heterophoria can exist as a cause of convulsive seizures independently of any extreme error of refraction.*

In case XXII it was deemed wise to give $+ 1.00^s$ glass for constant wear; but in Case XXIII almost absolute emmetropia was found.

I do not personally consider that any ground for argument exists to-day regarding the claim heretofore made by some oculists of repute,—viz., " that all heterophoria is due solely to errors of refraction." If such a statement is to-day believed by any one who has carefully investigated heterophoria, I think that Cases I, VI, XIII. XIV, XV, XVII, XXIII, XXIV, and XXV ought to disprove it.

Case XXIV was attended with outbreaks of apparent epileptic insanity, in addition to periodical attacks of continuous epileptic seizures of so violent a form as to justify the use of chloroform. Very frequent attacks of petit mal would also occur in spite of large doses of bromide salts.

The refractive errors were insignificant in this case, but the latent heterophoria was extreme. The case is incorporated chiefly to illustrate this point, as the eye-treatment was abandoned before any beneficial results could reasonably be expected.

Case XXV again illustrates extreme latent heterophoria, independent of any marked error of refraction. The results in this case cannot as yet be regarded as final. Some latent heterophoria still manifests itself at times. In spite of the efforts of the family physician, the parents discontinued the eye-treatment of this girl.

Prior to reporting the detailed histories of twenty-six cases of epilepsy I present a table that may possibly shed some light upon the eye-conditions[1] that are apt to be encountered in patients afflicted with this malady.

---

[1] The meaning of the terms used in this table and in the histories that follow have been given on page 35.

SUMMARY OF THE EYE-DEFECTS DETECTED IN THE CASES REPORTED.

	NO. OF CASES.	PER-CENT.	REMARKS.
Esophoria. . . .	24	93	Corrected in every case by graduated tenotomies upon the interni or advancements of the externi.
Exophoria	2	8	Corrected in every case by graduated tenotomies upon the externi.
Hyperphoria . . .	8	31	Corrected in every case by graduated tenotomies.
Heterophoria . . .	26	100	Is rarely absent in epileptic subjects.
Orthophoria	.	.	I regard this as an extremely rare condition in epileptics.
Hypermetropia	20	77	Corrected in most cases by glasses for constant wear.
Myopia	2	8	Corrected by glasses if in excess of two diopters.
Myopic astigmatism .	3	12	Fully corrected by glasses in every case.
Hypermetropic astigmatism .	7	27	Fully corrected by glasses in every case.
Emmetropia.	4	16	A somewhat rare condition, even in health.[1]

At a late meeting of the Ophthalmological Society of the United Kingdom a report of the refractive conditions of epileptics was made by Mr. Dodd (based upon the examinations of 100 epileptic subjects as contrasted with similar tests upon 50 healthy persons). The effects of treatment by a correction of the refraction alone by glasses were also reported.

Out of 100 epileptic subjects 75 were found to require glasses and were directed to wear them. Only 52 of these patients were faithful in obeying these instructions. Of these 52, 13 had no attacks for an interval of from one to two months, 36 were more or less benefited, and 3 were not apparently influenced by the correction of the refraction.

The conclusions drawn by Mr. Dodd were: (1) that *errors of refraction may excite epileptic seizures*, (2) that the *correction of refraction is of value in the treatment of epilepsy*, and (3) that *other sources of irritation may be suspected when a correction of refraction fails to afford marked relief.*[2]

To any one familiar with the later methods of investigation and treatment of heterophoria the *investigations of Mr. Dodd must seem singularly incomplete.* Yet it is interesting to note

[1] In this collection of cases the percentage of emmetropia appears to be greatly in excess of what is usually encountered.
[2] Italics my own.

that the correction of refraction alone accomplished enough in cases of chronic epilepsy to warrant a verdict in favor of the eye-treatment.

I admit that my deductions regarding the eye-tests in epileptic subjects derived from twenty-six cases cannot be regarded as final ; still I find that the preceding table indicates what to my mind would be about the average results of my personal examinations and treatment of quite a large number of epileptics during the past twelve years (a period when I first began to record, in every case, the results of examinations of eyes).

Because of the hopelessness of cure of chronic epileptics by drugs, and the importance of establishing beyond criticism the permanency of results in any new form of treatment, it may not be difficult for the reader to understand that an earnest seeker after truth is obliged to wait some years before he can himself feel sure that the beneficial effects of a treatment directed to a removal of a reflex cause are not temporary and misleading.

Before I give in detail the full histories of these cases I deem it not only wise, but absolutely imperative, to make some pertinent remarks relative to what causes may lead to a recurrence of epileptic seizures after eye-treatment ; these should in no way reflect upon the previous results obtained, nor should they be justly construed as clinical evidences derogatory to the method of treatment itself.

1. As the refraction of the eye should always be considered an important factor in epilepsy, any neglect on the part of the patient to wear the glasses prescribed, or any change in the formula, is apt to lead to a recurrence of the epileptic seizures.

One of the worst epileptic seizures ever encountered once occurred in a patient (who had been free from epilepsy for over a year) within an hour from the time that a strong cylindrical glass fell from the frame and was replaced with the wrong side out.

The removal of a spherical glass for a day (ordered by me for constant wear) caused an epileptic seizure. Case XI may be one of this type ; also Case III (see table published farther on).

Patients often break their glasses or frames and unconsciously get a wrong glass put in by some inexperienced or incompetent optician. They sometimes get the proper glass improperly inserted by some local jeweler or incompetent optician after mending the frame, etc.   Such occurrences are not infrequent, and a return of epileptic seizures is particularly apt to follow.

When patients are instructed to wear strong cylindrical glasses constantly, a simple bending of the frames may throw the glasses so much off axis as to create a far greater eye-strain than an actual omission to wear them. Case V in the table I believe to be an illustration of this type of accident on some occasions, although he has at times caused an epileptic seizure by excessive overloading of his stomach with wine and indigestible food.

It is needless to multiply illustrations relative to this point. The greatest care and fidelity on the part of a patient, combined with intelligence and education, can alone prevent the possibility of an occasional epileptic seizure from imperfectly corrected refraction after the case is dismissed from the care of the oculist. Among ladies, who generally insist, from pardonable vanity, in wearing eye-glasses instead of spectacles, the bending of the nose-clips and spring is always liable to distort the proper adjustment of the glasses, and to cause, unconsciously, a new source of eye-strain to the patient.

It can be easily understood by the reader that what originally induced epilepsy may tend to cause its recurrence, and that neither the oculist nor the method is responsible for accidents that he cannot avoid or the patients always be aware of.

Suppose a victim of malaria should be restored to perfect health by drugs, but, after a renewed and prolonged exposure to malarial germs, some evidences should appear of a return of the old malady, would it in any way reflect upon the results obtained by drugs?

2. Many chronic epileptics unquestionably (in my opinion) are influenced (even after long intervals of relief from seizures) by their former "epileptic habit."

By this I mean that under conditions of extreme nervous weakness or disturbance, such as often follow severe indigestion, anxiety, fright, loss of sleep, excessive exertion, etc., these patients are peculiarly apt to have a convulsion, where ordinary patients would have simply a headache or some milder evidence of physical depression.

Such attacks, as a rule, mean nothing.   They are not to

be construed as precursors of a return of the old epileptic condition. They pass without causing much, if any, constitutional depression, and the patient goes on (as before) free from subsequent seizures until some similar exciting cause occurs to induce one (Cases II, XXI, and XXVI, I think, illustrate this point).

3. During the treatment of heterophoria (which in epileptics is almost universally latent) the oculist never knows positively until a year or two has elapsed whether he has established an absolute and permanent orthophoria. So long as any latent muscular trouble remains, occasional epileptic seizures are to be expected, even with the greatest care on the part of the oculist as to the perfect correction of the refraction of the patient and the greatest fidelity on the part of the patient as to following the directions of the oculist.

In a happy way some one may, before the millennium, discover a drug or other process that will enable the oculist to determine at one sitting all latent heterophoria that exists in any case, as we now are enabled to measure, while a patient is under the effects of atropine, all latent errors of refraction at one sitting; but, until that time, we will have to allow the patient to disclose it by piecemeal (as it were), and to patiently wait until, by proper scientific aids, we can feel sure that we are interpreting the eye-tests of the patient intelligently, and relieving the burdens as fast as they are disclosed by the patient.

Prior to the discovery of atropine this was the way that latent hypermetropia was treated: Glasses were given to the patient as strong as he would tolerate at first; and, gradually, their strength was increased by the oculist, as fast as the patient would tolerate the increase, until the full correction of the latent hypermetropia was apparently reached.

4. It is important in all cases of chronic epilepsy, while eye-treatment is going on, to be sure, if possible, that no other reflex cause exists to keep up the epileptic seizures.

While it is my custom to have the teeth, ears, and nose of almost all of my patients examined by experts, and the pelvic organs of many female patients carefully looked into by a gynæcologist, with the view of eliminating all reflex factors that may co-exist with eye-strain, it is not always possible to state that the removal of eye-strain alone in unsuccessful cases has completed the treatment, nor is it fair to infer that the eye-work has been unproductive of any benefit.

A general proposition regarding the treatment of epilepsy may be thus stated: Every reflex cause that can be detected in an epileptic ought to be removed. The amelioration of the epilepsy may not come at once, and it will not come at all in a small proportion of cases; but an effort should be made in each case to give nature every possible chance to re-assert herself by relieving the nerve-centers of all sources of reflex irritation.

Many patients either neglect instructions to wear the glasses prescribed to correct errors of refraction constantly, or from pride, imaginary discomfort, the advice of others, or equally trivial reasons they discard them altogether (sometimes after obtaining most satisfactory results from their use).

It is not uncommon, in the author's experience, to find eye-glass- or spectacle- frames so distorted by bending of the clips, the nose-bridge, or side-bows as *to throw cylindrical glasses decidedly off from the proper axis*, thus subjecting the patient to a strain (not suspected, as a rule) that is even greater than would occur if the cylindrical glasses had been entirely discarded.

Again, whenever frames become accidentally broken and are repaired by some local jeweler or incompetent workman, it is not uncommon to find that the frames have been made too wide or too narrow for the patient, and the glasses decentered, reversed, or misplaced. The author has seen some very serious relapses of epilepsy follow such an experience, and it is now his custom to caution patients about the risks of not having glasses properly mended and frequently tested to see if they correspond with the prescribed formula. It is also his custom to have each patient supplied with duplicate glasses, to wear in case of accident while the damage is being repaired by competent workmen.

I know of one patient who, after wearing a hyperphorial prism of 4°, was persuaded by the lady he was soon to marry to discard his glasses while he attended a wedding-reception. He fainted, on returning home, in the elevator of his hotel, and was seriously ill. An epileptic patient, who had gone over a year without any convulsive seizures after eye-treatment, had a very violent fit within an hour after a strong cylindrical glass had slipped out of her frame and was replaced improperly.

Changing this glass to the proper axis at once restored this patient to health and comfort.

It is useless to enumerate here case after case to illustrate what those referred to sufficiently indicate,—viz., the *extreme care that the patient must take* (as well as the oculist) in correcting eye-strain for the relief of nervous diseases.

Many a case of eye-strain is permanently debarred (in the experience of the author) from ultimate recovery by acts on the part of a patient over which the physician exerts no control.

Let me cite a common example : A patient passes from the hands of a skillful oculist practically well; he remains well for quite a long period of time ; suddenly a relapse of the old symptoms occurs without any apparent cause, and the patient is told by some physician or friend to abandon eye-treatment because of this relapse. Now, the cause of this relapse may be a misplaced or decentered glass; some latent heterophoria uncorrected ; spasm of the accommodation (causing a $+$ glass to be a source of disturbance rather than of relief); possibly a bent spectacle-frame or some other condition that the oculist could easily rectify.

The more ignorant the patient, the less likelihood of restoration to health by eye-treatment,—especially when serious errors of refraction complicate heterophoria. Even some otherwise intelligent physicians seem to expect that a patient once restored to health by the relief of eye-strain must remain so indefinitely, irrespective of whether they eat or drink imprudently, whether they wear their glasses or not, whether their frames are wide, narrow, or bent, or whether they use their eyes under favorable or unfavorable conditions. Such critics might, with equal justice, deny the cure of a victim to typical malaria until he had lived for years after recovery in a malarial region without drugs and without a return of his former condition.

CASE I.—Mr. B., aged 26 years; machinist; married. Patient was referred to me by Dr. E. W. Hedges, of Plainfield, N. J., on July 30, 1892.

Family History : No hereditary tendencies to nervous troubles. Mother has headaches. One uncle had phthisis.

History of the Case : Was perfectly well up to the age of 24. Had an attack of petit mal with "spitting of blood." Had five attacks during the next twelve months. They then began

to be frequent, and within the next year six or seven attacks
daily were liable to occur. Bromides were begun two months
ago and have given little or no relief. From two to ten attacks
daily now occur. The patient says that he "does not fall in
these attacks, but twitches some and becomes stiff." He is
only momentarily unconscious and often can talk. His eyes
feel very queer during the attack, the pupils dilate, and the
conjunctiva is often inflamed after the attack.

This patient had a number of attacks in my office; so that
both my assistant (Dr. W. R. Broughton) and myself had ample
opportunity to observe and record the clinical features of several
of these seizures. Once, while walking from my reception-room
to my consultation-room, he suddenly stopped; his face became
slightly livid, with a fixed stare; his head was twisted to one
side; his fingers and arms worked convulsively; saliva drooled
from his mouth; he muttered incoherently as the attack was
passing off; and he urinated in his trousers. He was perfectly
unconscious for a period of about ten seconds. While some of
his attacks were somewhat lighter than this, he almost always
urinated in his clothing.

Patient does not smoke, nor drink anything but beer with
his meals. Has never had any venereal disease. He sleeps
well and has a good appetite. Has no marked tendency toward
constipation of the bowels. No history of an exciting cause
for the seizures can be elicited. Urine normal.

Eye-defects: At the first examination an apparent slight
hypermetropic astigmatism, but under atropine a *latent hypermetropia* of only $\frac{1}{2}$ diopter. Apparent esophoria of 1° at first
visit, but within a few weeks 8° of esophoria disclosed itself,
with a marked outward jump when the exclusion-test was
employed. With the Maddox cylinder-test: Esophoria of 10°;
adduction, 33°; abduction, 2°; sursumduction, right, 2°; left,
2°; no hyperphoria. This patient, on cessation of bromides,
had thirty-nine attacks during first week and sixty-seven attacks
during second week. During part of this time he was wearing
prismatic glasses for the purpose of developing the latent
esophoria that disclosed itself with great rapidity.

Treatment and Results: Three graduated tenotomies have
been performed upon this patient for the relief of his esophoria,
as it disclosed itself from time to time. From the first operation
the attacks became less frequent and the patient greatly improved

in his general buoyancy of spirits. The last record of any eye-tests of this patient was made on April 15, 1893. He had been entirely free from any attacks for thirteen weeks. His adduction was 33°; abduction, 7°; sursumduction, right, 2° +; left, 2° +; no hyperphoria; esophoria, 0.75°.

A letter from him, dated October 6, 1894, makes no mention of any attack up to that date,—*a period of over ten months without seizures, and with a total cessation from all drugs.* A subsequent letter, dated November 19, 1893, says: " With the greatest of pleasure I can say that I have not, since you saw me last, had any attacks or even the slightest symptom of any."

This patient has himself reported frequently to Dr. E. W. Hedges, of Plainfield, N. J. Concerning him, Dr. Hedges writes me as follows :—

July 7, 1896.

Dear Dr. Ranney: Replying to your letter of July 6th, in which you make inquiry as to the present condition of Mr. B. and Mrs. G.,[1] both of whose eyes you operated on for the cure of epilepsy, I can only report fully in regard to one.

Mr. B. has been perfectly well ever since. He works every day and has not had a single convulsion nor anything like one since you discharged him.

Mrs. G. moved to Buffalo about two years ago, since which time I have heard nothing as to her condition. Just previous to leaving she called upon me and declared that she was a different woman, mentally and physically, since the operation; that it seemed to her as though she had been living in a dream for years past. She had gained about thirty pounds in weight, as I remember it, and certainly looked younger and better than I had ever seen her. At that time there had been no return whatever of the epilepsy.

I watched these two cases for years, and saw them grow steadily worse under various forms of treatment, both dietetic and medicinal, and I am convinced beyond doubt that at least some cases of epilepsy can be cured by proper operations done upon the eyes. I am yours truly,

Ellis W. Hedges.

My critics may say that this case is one of petit mal, which is true ! But is not petit mal regarded, by most standard authors on nervous diseases, as not only a type of " genuine " epilepsy, but also as one of the most intractable types of epilepsy? Dr. M. Allen Starr says, on page 273 of his work[2]: " The treatment of petit mal is less satisfactory than that of grand mal. The only remedy of any service is nitroglycerin." In spite,

---

[1] See report on Case XII and letter.
[2] Familiar Forms of Nervous Diseases.

therefore, of the unmistakable character of these attacks and the failure to employ nitroglycerin as a curative agent, this patient made a quick and permanent recovery after the correction by me of his eye-strain. The evidence is overwhelming! The letter of the patient (1894) has been given; the written testimony of the doctor and my own records are also published here. Prior to my treatment the records show that from two to ten attacks occurred daily. This patient had so slight a refractive error (see table) that no glasses were prescribed.

CASE II.—Mr. F., aged 27 years; single; accountant. Patient was referred to me by Dr. J. B. Bissell, of New York, on September 2, 1892.

Family History: All mother's family died of phthisis except one. All maternal relatives have headaches and are very nervous.

History of the Case: Patient has always been exemplary in his habits,—neither smokes nor drinks and goes to bed at regular hours. Until the age of 12 years the patient was apparently well and strong. During that year he fell from the woodshed on to the ice and had a hæmorrhage of nearly a quart from the lungs. An epileptic convulsion followed. At intervals, for a long time, during the night he would have peculiar attacks associated with a peculiarly hard breathing and slight tremor. He has always been a victim to headache. These at times were terribly severe.

From the date of his first convulsion (fifteen years ago) this patient was placed upon the bromide treatment, which he has kept up ever since with varying doses. He has taken chiefly a combination of bromide of sodium and ammonium. He has been a chronic dyspeptic for years. No evidence of depression or injury to the skull exists. In spite of bromides, severe epileptic seizures would occasionally occur. During the first seven months of 1892 patient had four epileptic seizures at intervals of about two months apart.

He consulted Prof. E. G. Janeway, of New York, in the summer of 1892, who advised him "to see an oculist and continue his bromides." Weak myopic cylinders were then prescribed by Dr. St. John, of Hartford, Conn., which he has worn constantly for the past few months. His dyspeptic troubles seemed to be much lessened by the use of glasses, but an epileptic attack followed in about five weeks in spite of his bromides.

Eye-defects: Under atropine this patient disclosed the following refractive condition: O. D. V., $\frac{20}{40}$, with + 0.50^c (axis, 90°); O. S. V., $\frac{20}{40}$, with —0.50˙$\bigcirc$—0.50^c (axis, 180°). He also showed left hyperphoria, 1°; esophoria, 2°; adduction, 11°; abduction, 5°—; sursumduction, right, 2° +; left, 3°.

On the following day the effect of esophorial prisms was tried upon this patient. He showed within three hours unconquerable homonymous diplopia (with a red glass) and 6° of esophoria. Abduction, 0.

Treatment and Results: Graduated tenotomies were performed upon both interni within a space of two weeks, after which he showed left hyperphoria, 0.5°; esophoria, 0.75°; adduction, 21°; abduction, 7°; sursumduction, right. 2°; left, 3°.

Six weeks afterward he reported that his digestion was very much improved, that he had had no attacks for six weeks and taken no bromide for four weeks. He was wearing constantly a full correction for his refraction. Later on an investigation of his hyperphoria disclosed 4° of latent left hyperphoria and a 3° prism was incorporated in his glass. At this time he showed: Esophoria, 0.5°; adduction, 18°; abduction, 5° +; sursumduction, right, 0; left, 6°.

When four months had passed without an attack, the patient again visited my office. He reported that all evidence of chronic dyspepsia had absolutely disappeared and that he no longer had any distress in his head. If his glasses are removed he has unconquerable vertical diplopia. During all this period he had used his eyes for book-keeping throughout the entire day. His eye-tests showed: Left hyperphoria, 4.5°; esophoria, 0.5°; adduction, 17°; abduction, 5° +; sursumduction, right, 0; left, 8°—. The right inferior rectus was now operated upon for the correction of the hyperphoria. After this operation the patient showed: Left hyperphoria, 0.5°; esophoria, 0.5°. Three days later there were: Adduction, 23°; abduction, 5° +; sursumduction, right, 3°—; left, 3°—; left hyperphoria, 0.25°; esophoria, 0.5°.

A letter, dated March 13. 1893, says: "Saturday will be *seven months since the last attack,* and I am feeling simply immense."

On April 17, 1893, the following muscular tests were recorded: Vision $\frac{20}{20}$ in each eye with the glasses prescribed; no vertical or lateral muscular defect; adduction, 32°; abduction, 7° +; sursumduction, right, 2°; left, 2° +. Since the last note the patient reports that on July 23, 1893, he had a very slight seizure on going to bed,—the first and only one he has had since coming under my care.

It is important for the reader to note that an unfavorable prognosis regarding marked amelioration or cure had been given this patient by Prof. E. G. Janeway early in 1892, who referred him to an oculist and advised the continuance of bromides. He was then having typical attacks of grand mal on an average of one every eight weeks, in spite of the bromides. A letter received from the patient (October 17, 1896), in answer to one of inquiry from me, states that eight attacks have occurred in five years. This is about one-quarter of his average when under bromides. He also says: "My health since March 7, 1896, has never been better, and, in fact, couldn't be better."

I regard this case as one of practical cure, although occasional epileptic seizures have occurred. The patient is no longer afflicted with the severe dyspeptic troubles that persisted until I treated him, and his physical and mental condition to-day makes a marked contrast with that which existed when he was a victim to bromides. Unfortunately, the results in this case, that might otherwise have been expected, have been delayed by his occupation. He has to work, of late, long hours as a book-keeper, and to wear over one eye a hypermetropic cylinder and over the other eye a combination of a myopic glass and a myopic cylinder. He has also been obliged to use his eyes almost immediately after each operative step, and systematic out-of-door exercise is often rendered impossible by his business. He has once passed eight months without a seizure; again, nine months; and, again, eleven months. I feel personally convinced that he would remain entirely free from attacks if he lived out-of-doors and had an occupation in which he did not have to use his eyes. He is a strong, hearty eater, and needs systematic exercise to keep him in good physical shape.

Case III.—Master G., aged 10 years. Referred to me by Dr. A. D. Stewart, of Port Byron, N. Y., on May 5, 1893.

Family History: The father came with the boy and was

a picture of robust health. No epilepsy had appeared in his ancestry.

History of the Case: This boy came to my office accompanied by his father and Dr. Stewart, who led him along by the hand as he would a child of much younger years. He had a peculiar vacant, semi-idiotic expression that strongly indicated his enfeebled mental condition. He had one of those peculiar heads that was nearly spherical, with eyes set wide apart, a small nose, and mouth half-open. He was singularly sluggish both in his mind and in his movements, and was justly regarded by the doctor and by his parent as partially idiotic. His father was a farmer, and the boy had been brought up on the farm.

The clinical history, as well as the family history, was crude and imperfect. The boy had had innumerable epileptic fits, which nothing had seemed to relieve. On his way from Elmira, N. Y., to New York City he had had *several attacks on the train*, and had sat throughout the trip without hardly moving from his seat, except when in convulsion. He evinced no interest in any of his surroundings, did not look out of the window at the landscape, and in all other ways manifested his defective mental condition.

Eye-defects: A casual glance at the patient showed that he had an approach to double convergent strabismus. Vision, $\frac{20}{20}$ in each eye before atropine was instilled. Under homatropine a latent hypermetropia of + 1.75 diopters. On exclusion of either eye a marked jump outward confirmed the existence of esophoria of a high degree. All attempts to get accurate muscular tests proved futile, as the patient suppressed images in spite of all my efforts to make him show his ocular maladjustment. As nearly as I could find, 10° of manifest esophoria, at least, existed.

A free graduated tenotomy was performed upon both interni on the same day. with the effect of causing a remarkable change in the looks of the boy's eyes. A full correction of the hypermetropia (+ 1.75ˢ) was ordered for constant wear, in strong spectacle-frames, and an atropine solution ordered for instillation into the eyes. On the following day I succeeded in getting the following tests with fair accuracy, the suppression of images having been partially overcome: Esophoria, 5°; left hyperphoria, 1°; abduction. 6° +. *No jump on exclusion of either eye.*

The father and Dr. Stewart reported that, on the day of the

operations (within two hours following them). the boy had amused himself all the afternoon counting pennies and playing with marbles; that a remarkable change seemed to take place almost immediately in his mental apathy; and that he had not seemed as bright in many months. Within the next few days the boy's mental condition changed still more wonderfully for the better. He would enter my office with a laughing face; walk up to me, and shake hands without being told to do so; would talk to me about things he had seen; and showed the greatest interest concerning various instruments, etc., in my office. He had no convulsions for several consecutive days.

As I felt sure that (with the disappearance of atropine) the strong spherical glasses would be apt to cause the boy serious discomfort, and as no immediate prospect for further operative work seemed probable, I advised Dr. Stewart to take the boy home and to keep up instillation of atropine in the eyes at intervals, whenever the glasses caused the boy any discomfort. I also gave him a letter to an oculist near by in order that a specialist might make tests of the boy's eyes from time to time and inform me of his progress.

On the journey home. a report states, "the boy was so frolicsome and inquisitive on the train that he required constant supervision and care." The contrast was so great with his demeanor during his trip to New York as to excite the greatest surprise. No convulsions occurred during the trip. A most pathetic scene occurred at the depot on the arrival of this boy, his mother becoming hysterical over what she termed "his resurrection to health and mind."

The following letter was received on August 14, 1893, from the oculist to whom he was referred by me:—

DEAR DR. RANNEY: You will be glad to hear that your patient, R. G., of Port Byron, *has had no return of his epileptic attacks* so far. His father reports him much less restless and better tempered. At the end of June, after the use of atropia for a week, I found 2 50 D. of hypermetropia, and consequently increased the strength of his glasses to + 2.25 D. With these his V. = ⅔; there is esophoria of 2° or 3°, certainly not more, and abduction of 8°. The boy keeps his mouth open a good deal of the time, and I think possibly this and his decidedly "weak" expression may be due to post-nasal adenoids. I propose to have this point settled when he comes to me next, which will be at the end of September. In the meantime there is no question as to the great improvement in his condition following immediately on the correction of his refraction and strabismus.

I am yours very sincerely,

F. W. MARLOW.

During a personal interview with Dr. Stewart (October, 1893) he said that the boy was working steadily on his father's farm,—a source of income. instead of a constant care, as formerly. He also said that the father was very negligent in carrying out the instructions to have his boy wear the glasses constantly and have his eyes watched regularly, apparently being too ignorant to realize the importance of the matter, the gravity of the case, or the difficulties encountered in the treatment of epilepsy even under the most favorable auspices. As far as Dr. Stewart then knew. no epileptic seizures had occurred, but he expressed a fear that the negligence of the parents might create further seizures.

Late in 1896 I investigated the progress of this case and found that his parents had totally disregarded my instructions about keeping the glasses prescribed by me upon the boy; and, in spite of protests from Dr. Stewart and the oculist to whom I referred them, they allowed him to relapse into epileptic idiocy.

I quote the following letter from Dr. Stewart, written in answer to my letter of inquiry about this patient:—

PORT BYRON. N. Y., July 9, 1896.

DEAR DR. RANNEY: Yours of the 7th inst., asking about the condition of R. G., is received. When I last saw him, a few months ago. he was having fits worse than ever, and was fast becoming idiotic.

I received, about a week ago, a letter from Dr. ———, desiring information regarding the results of your treatment of the case.

I gave him a full history of the case, laying particular stress upon the remarkable change in the boy after the operations and the use of glasses; the entire freedom from spasms for six weeks; in fact, his general improvement until he lost his glasses in the river.

I told him of the parents' refusal to return for further treatment, and added that I had not the least doubt that the boy would have been cured had he continued under your care.

With kind regards, I remain, respectfully yours,

A. D. STEWART.

The report of this remarkable case has been given quite fully by me, and my only regret is that I cannot add this case to my reported cures, as I had every reason to hope I could do, under further inspection and treatment.

This case illustrates the extreme difficulties that are met when ignorant people undertake a class of scientific work that requires time, patience, and skill to get satisfactory results.

This case was, at best, a desperate one; yet in six weeks more had been accomplished than even the physician or family had reason to hope would ever be done.

In spite of this, the parents expected that the fits and idiocy would cease immediately, and they refused to take any further steps or to follow instructions regarding even the wearing of glasses, after the first fit occurred.

CASE IV.—Mr. H., aged 43 years; merchant. Patient began to have severe epileptic fits when 17 years of age. Had masturbated when a boy, and had been addicted in later years to excessive venery.

Family History: One brother is a confirmed dipsomaniac; the father died of paralysis; one sister is a victim to sick head aches; no phthisis has existed in the family, so far as could be ascertained.

History of the Case: The epileptic seizures of this patient varied in frequency from two or three a week to one in three months. He came under my care in 1871 (when 28 years old), and was treated by me for many years with enormous doses of the bromides of potassium and sodium. These salts reduced the attacks to about four a year. Stopping the bromides invariably increased the frequency of the attacks. Subsequently he spent a small fortune in undergoing various forms of treatment (dietetic, electrical, and medicinal) without any beneficial effects upon his epileptic seizures.

Eye-defects: In January, 1886, his eyes were examined after his return from an extended residence in the South. He showed, under atropine, a *latent hypermetropia* of 2.50 D., and also a manifest *esophoria* of 4°. Subsequently several degrees of latent esophoria also became manifest. Abduction, 4°; adduction, 30°.

Treatment and Results: Partial tenotomies were performed upon both interni, and hypermetropic glasses (2.50 D.) were given him. Since the first operation (January, 1886) he has taken no bromides and has not had a convulsion. He has twice been at "death's door" with fevers, but he has shown at no time any epileptic tendencies. I have had an opportunity of examining the eyes of this patient within the past month. At the last test he showed an abduction of 8°, with $\frac{20}{20}$ vision in each eye, and an adduction of 45°. He was wearing his glasses constantly when I last saw him (about two weeks since), and had been perfectly well for some eleven years. He said that he considered his epilepsy a thing of the past.

I would state that the attacks of this patient were typical grand-mal seizures, with lividity, complete unconsciousness, frothing, biting of tongue, and severe convulsions. He once worked himself under a sofa while in a fit. He has been carried out of a theater while in convulsions. Between the attacks he was perfectly well, and was regarded as one of the brightest speculators on the street. This was originally one of the most hopeless of chronic cases of epilepsy, and a complete and permanent cure was more than any one would have been justified in expecting, at the beginning of eye-treatment.

It demonstrates most conclusively, to my mind, the fact that *chronic epilepsy does not necessarily indicate any organic changes in the nerve-centers.*

CASE V.—Mr. H., aged 24 years; single; manufacturer. Referred to me by Dr. E. L. Mellus, of Worcester, Mass., on March 18, 1890.

Family History : Father has a nervous temperament; mother has gout severely and defective eyes; paternal grandfather died of bowel-trouble; paternal grandmother died suddenly from some unknown cause; all paternal uncles and aunts, seven in number, lived to be from 75 to 90 years of age; maternal grandmother died of phthisis and insanity (after childbirth), and was nineteen years in an asylum; one of her sisters died of phthisis; one maternal aunt died of phthisis at 16 years of age; maternal grandfather had gout terribly, drank heavily, and died of paralysis.

Eye-defects : This patient showed an enormous amount of unilateral astigmatism in addition to marked heterophoria. Right eye, $+ 0.50''$ (axis, $90°$); left eye, $+ 4.00°$ (axis, $180°$) $\bigcirc - 1.00°$ (axis, $90°$); esophoria. $11°$; adduction, $35°$; abduction, $3°$; right sursumduction, $4°$; left sursumduction, $4°$.

History of the Case : Patient was perfectly healthy until he went to Exeter to fit for college. While in Exeter he had several epileptic seizures. He had no special aura, but usually bit his tongue. Had masturbated before his attacks, but has not since. He then entered Harvard and stayed a year. He had, he thinks, four attacks during that year, during which he took no medicine. He left Harvard in June, 1885, and in September, 1885, he went into the draughting-rooms of his father's factory. For eighteen months he took no medicine, and in that time had several attacks.

While in Cuba in 1887 he had a very severe attack and began taking bromides. Within two months, while in Mexico, he had two serious attacks, cutting his chin badly in one and knocking out a front tooth in another. He then came to New York and consulted an eminent neurologist, who increased his bromides and put him on restricted diet. He had only one severe attack and one of petit mal during the next year, but his mental condition became seriously impaired.

When he came to me his whole appearance and manner showed markedly the poisonous effects of the bromides. His face was covered with acne. His mental condition was so bad that an interested conversation was almost impossible. In fact, it had become so alarming that his father, with tears in his eyes, said: "Although he is my only son, I would rather see him dead than in his present condition."

Treatment and Results : All bromides were at once stopped. A full correction of his astigmatism was given for constant wear, and graduated tenotomies were performed on both internal recti for the relief of his esophoria. During the first six months of treatment, while his glasses were being changed and operations performed, he had five attacks,—four very light ones and one medium attack. All of these occurred after excessive indulgence in rich and indigestible food late at night, and one after indulging in too much alcohol. *During the next twenty months he had no attack of any kind.* He had been actively engaged in business and has gained twenty pounds in weight. He had regained perfectly his mental condition, traveled without an attendant, maintained a yacht, and was considered perfectly well by his parents and physician. Since June, 1892, patient has had no attack,—*a period of three years with only one slight seizure.* He has lately married, and has gained over twenty pounds in weight since he has abandoned bromides. His last tests showed the following conditions: Is wearing full correction for refractive errors; O. D. V. $= \frac{20}{20}$; O. S. V. $= \frac{20}{30}$; adduction, 38°; abduction, 6°; sursumduction, right, 4° —; left, 4° +; exophoria, 0.75°; left hyperphoria, 0.75°.

The character of the attacks of this patient cannot well be

questioned. He had cut his head badly in one fit and in another had knocked out a tooth. His father described his attacks, when he brought him to my office, as "frightful to witness"; and he gave in detail all the symptoms of grand-mal seizures. His heredity was a bad one. He had been for some time under the care of a leading neurologist of New York City, who had given a very unfavorable prognosis, and had steadily increased the doses of bromides until the mental condition of the patient was deplorable. He was not allowed to go about without an attendant when I first saw him.

I quote a letter from this patient, in answer to my inquiries, as follows :—

July 24, 1896.

MY DEAR DR. RANNEY: I have your letter of July 22d. In reply, I would say that I have had no recurrence of my old trouble since April, 1895

Sincerely yours,

H.

Dr. E. L. Mellus, who sent this patient to me, lately returned from a two years' tour of Europe—which he has devoted, I understand, chiefly to the study of ophthalmology, because his interest in this case and the treatment of his own eyes by me awakened him to the importance of this field. I have been unable, as yet, to get a written reply to my letters to him regarding this case, as I have not ascertained his present address. I regard this case as one of practical, if not absolute, cure from a most unfavorable beginning. The isolated seizures that he has had (three in nearly six years) have all been due to a marked gastric upset,—from gross imprudences[1] in eating, drinking, etc., combined with or due to some maladjustment of his glasses.

He is to-day in charge of one of the largest manufacturing industries in New England ; nearly two years have passed since any sign of epilepsy has appeared ; and, prior to the last attack, only one very slight seizure occurred in three years. He has married, and is far above the average man in intellect to-day. This patient is wearing an extremely strong and complicated glass over one eye, and the slightest bending of his spectacle-frame alone might cause, and I think has caused, at least one epileptic seizure, although he had done other things to upset his stomach. If this case only were all that I had to report, I

---

[1] Two of these attacks followed excessive use of champagne with lobster-salad.

should feel that the claim that eye-strain must be accepted as a cause of " genuine " epilepsy was established. It is a practical cure from a most unpropitious beginning. I could quote from a number of letters written by others regarding this patient,— many expressions of amazement and delight over the wonderful change that had followed the eye-treatment. His immediate family and friends regard him as cured. His eye-tests are apparently normal, and his attention has been so frequently called to the importance of keeping his spectacle-frames in their proper relationship to the eyes that a marked displacement of his strong cylindrical glass is not likely to occur in the future.

CASE VI.—Mr. S., aged 19 years. First seen by me on November 27, 1888.

Family History : Mother has frequent and severe sick headaches, and her sister is a martyr to them also. The brother and sister of the patient have headaches. The paternal heredity could not be accurately given by the mother. who brought the boy to me for treatment. No phthisical tendencies had ever manifested themselves in any of the patient's ancestry, as far as known. Every known relative on the maternal side suffers from headache. The mother had very marked esophoria. She was treated by graduated tenotomies in my office with very great benefit. The cure of her headaches resulted. They had made her life a burden for years.

History of the Case : The patient is a tall, finely developed young man of 5 feet 10 inches, weighing about 150 pounds, and with a good color. His mother gave the following facts : Up to the 14th year of age the patient was in perfect health. He then had his first epileptic seizure. following upon an attack of so-called " congestion of his brain," for which no cause could be found except a fall while skating. He was then at school in Paris. Within the next year, in spite of bromides, he had three " fainting attacks," lasting an hour each. He was then removed to a school in England, and had a number of severe epileptic seizures in spite of large doses of bromides. In August, 1887, he was sent to America and placed in a select school, where he could be carefully watched over and his habits of life regulated. He had taken every day for the previous year not less than 60 grains of potassium bromide and 15 grains of sodium bromide, and at times much larger doses.

During the year prior to his visit to my office the seizures had become more frequent and extremely violent,—so violent that three men could not restrain the patient, *and a room had been padded with mattresses and especially kept for the protection of the patient when thus seized.* Into this room he would be placed and allowed to thrash about, until attack after attack would prostrate the patient. All medical treatment seemed of no avail, and the father was asked to remove the boy from the school. Medical advice was then taken, and it was deemed advisable to commit the patient to an insane-asylum as an incurable and uncontrollable case of dangerous epileptic seizures.

At the earnest solicitation of friends the parents were urged to make a trial of the eye-treatment, in the hope that it might possibly avert so horrible a fate for the boy, even if it did not markedly affect the frequency of the fits.

As the absolute cessation of the use of bromides was insisted upon by myself, from the date of the first visit, the boy was, with some reluctance, admitted, at my solicitation, to a private hospital, so that he could come to my office with an attendant and be protected from injury if the fits became very frequent or severe. A record kept by the principal of the school showed that fits had occurred during thirty-four days, in the twelve months that preceded my care of the boy, in spite of extreme doses of bromide salts and chloral, at intervals, in addition to his regular daily doses.

Eye-defects: On the 17th of November, 1888, the patient showed normal vision in both eyes; adduction, 54°; abduction, 5°—; right sursumduction, 2°—; left sursumduction, 2°—; esophoria, 2°; left hyperphoria, 1°. In accommodation patient showed esophoria of 10°.

November 18th. Under atropine patient showed: Latent hypermetropia, ½ diopter; esophoria, 2°; left hyperphoria, 0.5°.

Treatment and Results: The patient carried his head, as his mother said he always had done, very markedly to the left side (justifying a suspicion that right hyperphoria actually existed), and his esophoria was very palpable to a careful observer. Later on my suspicion of an existing right hyperphoria became confirmed by most positive tests. Here, then, was a boy who showed, at the onset, almost perfectly constructed eyes, with only a slight tendency inward (apparently) and a suspicion of hyperphoria, yet he was having terrific epileptic seizures that were uncontrollable by drugs. His power of abduction was low, however, and prisms of 1°, base out, were placed for twenty-four hours over each eye. At the third visit, on the following day, he showed esophoria of 7°; adduction, 71°; abduction, 3°; and the prisms were increased to 4°. In three days more he showed esophoria of 10°, with unconquerable double images. Adduction, 68°, and abduction, 2°. The tendon of the right internal rectus muscle was freely relaxed by a graduated tenotomy. This improved his deviating tendency inward, but some esophoria still disclosed itself. Result: Esophoria, 2°; abduction, 7°; left hyperphoria, 0.25°. Subsequently a 2° esophorial prism was given the patient for constant wear. He disclosed at once 6° of esophoria, and his abduction fell to 4°; adduction, 73°.

On December 24th the opposite internal rectus was likewise operated upon, and his esophoria was apparently totally corrected for some time after the operation, his adduction being normal (8°). On December 27, 1888 (ten days after the cessation of bromides), the patient had a fainting attack in my office without tremor, but with a total loss of consciousness for some ten minutes. The habit of carrying the head toward the left shoulder had been persistent up to this time. An examination of the eyes disclosed a *right hyperphoria* (as was originally suspected) of quite a high degree.

From this date until February 12, 1889, patient was treated for hyperphoria by prismatic glasses, and *no return of epileptic seizures had occurred.* At this date the patient showed: Esophoria, 0; right hyperphoria, 2°; adduction, 73°; abduction, 6°; right sursumduction, 4° +; left sursumduction, 0. I then determined upon a third operation, and let out the right superior rectus as far as I deemed it wise to do so, although I apparently failed to correct his right hyperphoria by so doing. Results of operation: Esophoria, 0; left hyperphoria, 0.5°; right sursumduction, 3° —; left sursumduction, 2° +. Prismatic glasses, for a remaining tendency toward a right hyperphoria that soon disclosed itself, were again resorted to as a step toward correction of the existing eye-strain. The boy then returned to his former school. On July 1st, when the *boy had passed over seven months* without an epileptic seizure, I received a letter from his mother, from which I quote as follows: "I want to tell you how very grateful I feel for the great good you have done my boy. It is really wonderful how he has improved in health since he has been under your care. He writes me that he has not had a single attack since I left New York. This seems almost a miracle, when one remembers how the boy suffered before coming to you."

By July 7, 1889, *over nine months* had passed since an actual convulsion had occurred, and nearly eight months since the "fainting attack" in my office. He had been some time without prisms or any eye-treatment, when he imprudently used a lawn-mower violently on a very warm day for several hours. As a consequence, he was seized with one of his "old-time attacks," having three severe convulsions and two light ones in the next forty-eight hours. They were accompanied by marked gastric disturbance and fever. On July 10, 1889, on visiting me, I found over 2° of right hyperphoria, for which I again operated upon the right superior rectus tendon. Tests after this operation showed: No hyperphoria; esophoria, 1°; abduction, 6°; sursumduction, right, 3°; left, 3° —.

On October 14, 1889, the patient engaged in a cross-country run of several miles, after school exercises, and became greatly overheated. He was again seized with a severe convulsion, and had two light ones later in the day. He had again marked gastric disturbance. Fourteen weeks had elapsed since the previous attack. An esophorial prism of 2° was given him.

On December 26, 1889, the patient was again seen. He had experienced no return of attacks, was in excellent health, and had taken no medicine for thirteen months. He still shows 1° of right hyperphoria; esophoria of 2°; adduction, 58°; abduction, 7° —; right sursumduction, 4°; left sursumduction, 1° +. He is wearing prismatic glasses for 2° of esophoria.

As operative work upon the eye-muscles had been done somewhat rapidly upon this patient, he was allowed to wear his esophorial prism, with instructions to visit the office at intervals of a few days, with the object of exercising the eye-muscles by means of prisms and improving their tone. Little change occurred. The old tendency toward right hyperphoria has never been disclosed, and gradually the right and left sursumduction have become equal. Some latent esophoria disclosed itself, however, and the abduction showed a marked decrease.

On October 29, 1890, the patient showed the following tests: Esophoria, 3°; left hyperphoria, 0.25°; adduction, 58°; abduction, 3°; sursumduction. right, 3° +; left, 3° —. He had been wearing a 2° esophorial prism. A graduated tenotomy was performed on the right internal rectus. Result of operation: Esophoria, 0; hyperphoria, 0; abduction, 8°.

By April 2, 1892, this patient had *passed over ten months without an epileptic seizure. and only one attack had occurred in nearly two years.* He is wearing now a 2° esophorial prism.

September 6, 1893. Patient reported at the office after an absence of several months. He reported that he had discarded the 2° esophorial prisms that had been ordered for constant wear, and had been without them for several days, during which time he had been occupied in a hot-house. He was seized with a convulsion that lasted some minutes. *Two years and three months had elapsed since his last epileptic seizure.* The tests of his eyes showed: Adduction, 45°; abduction, 5° +; sursumduction, right, 2°; left, 2°; esophoria, 2°; no hyperphoria; a 2° esophorial prism was ordered.

November 12, 1893. Since the last note the patient has had no epileptic seizure. His esophorial prism has been increased to 4°, as he has disclosed some latent esophoria. His last tests showed: Adduction, 69°; abduction, 3°; sursumduction, right, 2°; left, 2°; esophoria, 5°; no hyperphoria. Later the remaining esophoria was removed by a graduated tenotomy and his abduction carried to 8° +.

This patient departed for the West about January, 1895, and lived in a very high altitude. Soon afterward he reported that a series of three attacks had occurred after fainting from the pain caused by two enormous boils. These attacks should not be awarded any clinical significance.[1] He has had only this series of attacks in four years and six months. I attribute the relapse of his epileptic tendencies to living in a high altitude. Since his return to New York (nineteen months ago) he has never had a fit.

I regard this case as one of practical cure. Contrast the present condition of this patient with his past, when a room padded with mattresses was always kept ready for his confine-

---

[1] The reader is referred to previous remarks of mine (page 189) relative to causes that may induce occasional epileptic seizures in patients who have been victims to chronic epilepsy. The high altitude in which he was living appears to have acted badly upon his health.

ment, and thirty-four days of almost continuous convulsions occurred in one year. This patient was to have been committed to an asylum (as a hopeless and dangerous epileptic) on the sworn testimony of two medical men of repute had the parents not consented to try the eye-treatment at the solicitation of friends before taking so sad a step.

The "genuine" character of the epileptic seizures of this patient has never yet been called into question; yet, for the benefit of critics, it may be wise to state that all who have witnessed the attacks before I saw the patient concur in the description of a most horrible series of convulsions of extreme duration, with total unconsciousness, lividity of the face, frothing at the mouth, biting of the tongue, the epileptic cry, and a more or less prolonged stupor after each attack. Between these attacks he was as well as any person drugged with bromides and chloral could be. His mother had for years terrific and frequent attacks of typical sick headaches. She was cured of them later by graduated tenotomies performed by me upon her interni.

This patient comes into my office occasionally, and has been seen by me within a week. He is perfectly well, and for nineteen months he has had no sign of any epilepsy. In fact, not counting the three fits that occurred while he lived in an extremely high altitude, he has had no fits for over five years.

CASE VII.—Mr. S., aged 16 years; student.

Family History: Many cases of phthisis have occurred in maternal ancestry. All paternal ancestry are extremely nervous and excitable. The father is intemperate, extremely irascible, and often difficult to control; at times he has been considered partly insane. He has very marked eye-defects. A brother of the patient was an epileptic, partially idiotic, and an inmate of a school for the feeble-minded until he died.

History of the Case: First examination made on April 6, 1889. Patient has always suffered from an extremely weak stomach, and at intervals from peculiar attacks, supposed to be fainting attacks.

Eye-defects: Under atropine 1.00 D. of hypermetropia was found. While testing his eye-muscles this patient complained of great distress in his stomach, and was immediately seized with an epileptic attack in the office. He had had several such attacks (at varying intervals) while at a preparatory school. At the third visit he showed the following tests: Adduction, 33°; abduction, 3°; sursumduction, right, 1°; left, 2°; esophoria, 5° to 10°; left hyperphoria, 1° to 2°. In accommodation he showed an exophoria of 4°, and + 1.00⁸ glasses were ordered for reading. Esophorial prisms (1° each) were given for constant wear. At subsequent interviews it was found to be difficult to complete the muscular tests on account of distress in his head and stomach. He showed, however, an esophoria that fluctuated from 8° to 15°; adduction, 38°; abduction, 2°; sursumduction, right, 3°; left, 3°. Esophorial prisms (2° over each eye) were continued.

Treatment and Results: A graduated tenotomy was performed during November, 1889, on each internal rectus, and his prisms were removed. Six months after the last tenotomy the patient showed: Adduction, 25°; abduction, 6°; sursumduction, right, 2°; left, 2°; no hyperphoria; esophoria, 0 to 2°. In an interview with his mother in 1893 she stated that her son had not had a single epileptic attack to her knowledge since the operations on his eyes over two years and a half ago.

It should be noted by the reader that this patient had an heredity to epilepsy and insanity. One brother died of epileptic idiocy in an institution for the feeble-minded; the father had dipsomania, and, at times, had been regarded as insane. The fit that this patient had in my office was a typical attack of grand mal. He was totally unconscious, livid, frothed at the mouth, was rigid for some seconds, twisted his head to one side, then had clonic spasms of the limbs, and was drowsy after the fit passed off. He had to be held in a chair by my assistant and myself. He had, while at school, several similar attacks to this one prior to being placed under my care. Originally his attacks were milder and resembled a fainting-spell. Not long ago (during 1896) his mother called to have me examine another member of the family. She reported that her son was and had been entirely free from epileptic attacks.

CASE VIII.—Mr. O., aged 28 years; single; minister of the gospel. Referred to me by Dr. M. A. H. Hart, of Milton, N. H., on October 28, 1892.

Family History: No epilepsy or allied diseases among his ancestry.

History of the Case: The first paroxysm occurred while the patient was a student at the Andover Seminary after a hard evening's work at study. Altogether four typical epileptic attacks have occurred in addition to one or two of petit mal. *Three of these attacks have occurred during the past twelve months.* The last attack occurred while reading in a moving passenger-car. Patient had taken bromides regularly during the previous year, and was sent to me by his physician for the purpose of determining whether any reflex causes existed for the attacks. The patient appeared to be in good physical health and wore the following glasses, given to him by Dr. Lewis Dixon, of Boston: O. D., — 2.50° ◠ — 0.50° (axis, 180°); O. S., + 0.50° ◠ + 1.00° (axis, 130°). No epileptic attack had ever occurred until these glasses were given for constant wear.

Eye-defects: At the first examination the patient was found to have a *crossed diplopia of 15°*, with an apparent right hyperphoria of 2°. His abduction was 17°. Under atropine the following refractive condition was found: O. D. V., $\frac{15}{200}$, made $\frac{20}{40}$ by — 2.50° ◠ — 0.50″ (axis, 180°); O. S. V., $\frac{20}{40}$ —, made $\frac{20}{40}$ by + 0.50° ◠ + 0.50° (axis, 115°). With this correction the crossed diplopia was unconquerable when a red glass was placed before either eye. Without the red glass the following muscular tests were elicited: Adduction. 8°; abduction, 17°; sursumduction, right, 6°; left, 4°; right hyperphoria, 2°; exophoria, 16°.

Treatment and Results: With such extreme heterophoria, it was deemed wise to perform a free graduated tenotomy at once upon the right external rectus. After the operation patient disclosed: Exophoria, 6°; right hyperphoria, 3°; abduction, 12°. Subsequently a tenotomy for a right hyperphoria was performed, as the patient then disclosed 5° of right hyperphoria, with a sursumduction, right, of 8°; left, 1° +. Results after this operation: Right hyperphoria, 1°; exophoria, 5.5°; adduction, 23°; abduction, 14°. It was decided later to operate on the left external rectus. Results after this operation: Right hyperphoria. 0.5°; exophoria, 3°; adduction. 27°; abduction, 9°; sursumduction, right, 5°; left, 4°. In accommodation, esophoria, 4°, with full refractive correction. The patient now *showed no diplopia*, either vertical or lateral. with or without a red glass. Within the subsequent week the patient showed: No hyperphoria; exophoria, 1°; adduction. 27°; abduction, 9°; sursumduction. right, 5°; left, 5°.

It was a great surprise to me and at the same time a source of much gratification that in so short a time (about one month) an approach to ocular equilibrium in this patient had apparently been attained. It was, of course, possible and, perhaps, probable that some latent heterophoria still existed, which might require further investigation and correction; but I deemed it wise at this time to instruct the patient to desist as far as possible from reading and study for a few months and to live an out-of-door life. I recognized the fact that up to the time when glasses were first given him by Dr. Dixon, of Boston, this patient had *practically been a monocular being*, using his left eye chiefly and discarding or suppressing the images in his myopic eye. As soon as his myopia was corrected in the right eye the patient unconsciously was forced to use the two eyes simultaneously, and his extreme heterophoria became an active factor in producing reflex nervous troubles. I was hopeful that after orthophoria was nearly established this patient might tolerate a full correction of his refraction; but I suggested to the patient at the last interview that it might be well for him to discard his distance glasses in case epileptic seizures should occur. I tried to impress upon him the gravity of his malady, the necessity of careful observation of the frequency of his attacks, the length of time required for the complete solution of such cases as his own, and, above all, the importance of abstaining from bromides during the investigation of his ocular problems. I was greatly surprised, therefore, when I received a letter from this patient in November, 1893, in which he stated that he had abandoned the eye-treatment and returned to bromides.

CASE IX.—Miss S., aged 13 years. Referred to me by Dr. S. B. Casey, of Brooklyn, on May 28, 1888.

Family History: Mother is very nervous. One brother, aged 15 years, is very defective mentally. One paternal aunt died of phthisis.

History of the Case: Patient developed epilepsy at the age of 10. Prior to that age she had noticed visual disturbances of a peculiar kind. As she expressed it, "everything seemed to move downward." She had been an unusually bright child up to the time of the epileptic attacks. For some months the attacks steadily increased in frequency until they averaged one a week. She was then put under bromides in various combinations. In spite of the bromides and the extraction of five teeth, she continued to have two attacks a month, the two always following each other.

In April, 1888, at my suggestion, the experiment of discontinuing bromides for two weeks prior to the beginning of eye-treatment was contemplated, but within twenty-four hours a condition of continuous epilepsy developed, which seemed to threaten the life of the patient. It was found necessary at once to resume the bromides in large doses. A letter from the physician, dated May 5, 1888, says: "I must say freely that I am unable to promise more in this case than to modify the frequency and severity of the attacks."

Eye-defects: On May 25, 1888, the patient showed: Right hyperphoria, 0.75°; esophoria, 4° to 10°; adduction, 15°; abduction, 3°; sursumduction, right, 2° + ; left, 2° —. Under atropine: O. D. V., $\frac{20}{200}$, made $\frac{20}{20}$ — by + 1.50^s $\supset$ + 0.75^c (axis, 95°); O. S. V., $\frac{20}{100}$, made $\frac{20}{20}$ — by + 1.75^s $\supset$ + 0.75^c (axis, 115°). Subsequently a very large amount of latent muscular trouble was discovered.

Treatment and Results: A full correction for her refraction was ordered, atropine being instilled into the eyes whenever ciliary spasm manifested itself. Within two weeks a graduated tenötomy was performed upon both internal recti and all bromides were discontinued. The eye-tests showed: No hyperphoria; esophoria, 0 to 3°; adduction, 30°; abduction, 9°—; sursumduction, right 2° + ; left, 2° +-. No attacks for six days followed the cessation of bromides. All headaches stopped at once, and a marked drowsiness that accompanied the bromide treatment has entirely disappeared.

July 22, 1888. The following report was made: "No attacks have occurred for the past three weeks; bromides have been withdrawn for six weeks." The eye-tests remain unchanged from last note.

September 4, 1888. The spherical strength of the distance glass was decreased to + 1.00, the cylinder remaining the same, the reason for this change being a persistent tendency to ciliary spasm that exhibited itself at intervals.

October 4, 1888. Patient shows: Adduction, 28°; abduction, 11°; exophoria, 0 to 2°. She has had only two seizures in about five weeks, one of them preceding menstruation. No bromides have been administered during the past four months.

Her expression has undergone a very remarkable change for the better and her physical condition is very much improved.

A report on September 3. 1889, shows that *no severe attack has occurred since July 29. 1889*. Most of the slight seizures that have occurred have appeared during or just prior to menstruation. She has passed over four weeks without any seizure during this period. Her eye-tests on this date showed no hyperphoria; exophoria, 0 to 1°; adduction 27°; abduction, 8°; sursumduction. right. 2° ; left, 2°.

December 23, 1889. The patient reports: "I have had only five light attacks during the past ten weeks, in some of which I did not lose consciousness."

A letter. dated June 12. 1890, from a brother of the patient says: "It has occurred to me that you may have considered that ingratitude has been shown by a failure on my part to indicate how much my dear sister has improved under your advice and direction since the first visit made to your office. My sister has done well since her last visit to you. She is attacked monthly with one or two slight seizures, each decreasing in intensity until, according to her own statement, she merely felt the coming on and then it went away." This patient moved away from close proximity to New York some years ago; hence I have not had an opportunity of testing her eyes in that period. I have endeavored to get a report of her condition, but thus far have not been able to obtain her present address.

It should not be forgotten by the reader that this case was a young girl who had so many severe epileptic seizures within twenty-four hours after I advised the withdrawal of bromides that her physician despaired of her life, used chloroform, and hastily returned to the bromides. This fact illustrates forcibly that Dr. Peterson's published statement,[1] which I quote here, is not always correct. He says: "It is a fact. which has not as yet received sufficient attention, that in cases of chronic epilepsy long treated with bromides, relief from attacks for considerable periods of time follows diminution or cutting off of the bromides."

For nearly ten years I insisted that every epileptic patient who came under my care should pass at least one month without bromides or chloral. and keep an accurate record of their

---

attacks (severe, medium, or light) during this period. This withdrawal of all drugs always preceded any eye-treatment, and it was insisted upon by me because the basis of my records before and after eye-treatment (in my office) would thus be alike.—*i.e.*, the patient's condition would not be masked by the use of drugs for one month prior to the eye-treatment, and also while this scientific method of treatment was being tested.

Let me analyze the 26 cases (here discussed) from this stand-point. In only 12 cases was this point determined. Fits increased by withdrawal of bromides: Cases I. V, IX, XI, XVIII, XIX, and XX. Fits not modified or decreased by withdrawal of bromides: Cases VII, XIII, XIV, XXV, and XXVI. In the following cases the effect of withdrawal of the bromides upon the frequency of attacks was not determined: Cases II. III. IV, VI. VIII, X. XII. XV, XVI. XVII, XXI, XXII, XXIII, and XXIV. In the 12 cases where the effects of withdrawal of the bromides were determined by me, 7 experienced a marked increase of attacks and 5 experienced little. if any, modification of previous intervals between seizures. In no case were the seizures arrested for any length of time.

When patients have been drugged by all possible combinations of bromides. and often with chloral at the same time. until their mental faculties and physical powers have nearly reached their limit of endurance (and this is too often the case with patients sent to epileptic institutions). I do not wonder that the reports of hundreds of cases at Bielefeld and Craig Colony show an improvement by the withdrawal of the drugs from these poor victims. Anything would help them if it gave nature even a slight chance! This is the weak spot in all tabulated statistics regarding epileptics from institutions and clinics. The patients tabulated are too poor, too weak often in intellect. too imperfectly nourished. too heavily drugged in the past. sometimes too low in the moral scale, and generally too low in intellectual power to make such reports of as much actual value as they might appear numerically. Nothing can lie so much as figures ('tis said) when manipulated with skill; and cases taken from private practice of a higher type. with wealth. good home surroundings, more culture and intelligence, and who have had the benefit of good medical counsel in the past, are certainly a better basis for clinical deduction than the previous class described.

With this pardonable digression, which the history of this case brings to the surface, I shall proceed with some interesting facts regarding the treatment of Case IX. This girl was but a child when first seen by me. Her system had been saturated with bromides. She had not menstruated. She was very sluggish mentally, and physically she was extremely weak. She had an idiotic brother. As she was one of my earlier cases, I feel sure that her eye-problems were imperfectly solved by me. All tests for hyperphoria were then made with clumsy instruments, with a head-rest to insure immobility and a spirit-level on the frame of the prisms that rested upon the patient's nose while making the tests.

I am satisfied that the best results of eye-treatment were not obtained in her case, yet she showed a wonderful physical and mental improvement. Her fits were materially lessened in frequency. She took no bromides, and grew into a bright and attractive young lady. I have some very interesting photographs of this patient (taken at different stages of the treatment) which show very clearly the physical and mental improvement.

CASE X.—Mr. F., aged 40 years; married; teacher. Referred to me on October 22, 1890.

Family History : Mother living, aged 70 years. Father died at 63 years of age, probably of scrofula. One brother has had nervous prostration. No phthisis or nervous tendencies, so far as known, in any of the ancestry.

History of the Case : Six or seven years ago the patient began to have momentary lapses of consciousness with pallor, followed by sweating and confusion of mind. For the past two years *these have happened every five or six weeks.* A number are liable to occur at these periods in succession. These attacks are usually attended with nausea, would last about one minute, and would be accompanied by a peculiar dullness of perception and memory that would slowly disappear. The patient would be unfitted for work for from thirty-six to forty-eight hours, and several attacks would happen each day while he was so afflicted. Three weeks before I saw him, while driving, he suddenly became pale, stopped talking, gave a cry, and had a typical epileptic seizure. He was unconscious for half an hour. Patient has had internal hæmorrhoids to an exaggerated degree and some prolapse of the rectum. This patient had never taken any of the bromide salts.

Eye-defects: Without atropine, patient tolerated $+$ 0.50^s for distance. Under atropine. vision of each eye, $\frac{20}{70}$, made $\frac{20}{20}$ by $+$ 1.50^s. He also disclosed: Esophoria, 6°; left hyperphoria, 0.75°; adduction, 22°; abduction, 4°; sursumduction, right, 2° $+$; left, 2°—. With a red glass placed over either eye the *patient showed homonymous diplopia.* After wearing a 3° eso- phorial prism for a couple of days patient showed: No hyper- phoria; esophoria, 6°; adduction, 30°; abduction, 0.

Treatment and Results: On October 27, 1890, a graduated tenotomy was performed upon the right internal rectus. Re- sult: Apparent esophoria, 1°; abduction, 6°. An investigation of the esophoria by prisms during the ensuing week disclosed a high degree of latent esophoria (8°) with a decrease in abduc- tion (2°). A graduated tenotomy was then performed upon the left internal rectus. Result: Apparent esophoria, 0 to 4°; ab- duction, 6° $+$. Two weeks later (November 10, 1890) patient was called home. He showed: Esophoria, 1° $+$; no hyper- phoria; adduction, 34°; abduction, 6°; sursumduction, right, 2°—; left, 2°—. He was ordered $+$ 1.00^s glasses for reading and for constant wear as soon as they caused no annoyance, and to return later for subsequent investigation of the eye-con- ditions. I suspected at the time that the latent esophoria was still but partially relieved.

During the next year the patient was seen for about one week daily, during which time his ocular muscles were exer- cised by means of prisms. He disclosed (as I suspected that he would) some 6° of latent esophoria, and for part of the year he wore constantly $+$ 1.00^s $\bigcirc$ 1.5° prism, base out, over each eye. During this year he had one slight convulsive attack and five attacks of nausea.

During October, 1891, graduated tenotomies were per- formed on both interni, for the relief of the esophoria which he disclosed. He returned home three days after the last tenotomy, showing: Esophoria, 1°; abduction, 8°; adduction, 32°. The esophorial prisms were removed from his distance glasses.

January 19, 1892. Patient reports one attack of nausea lasting about a week. The spells of nausea during this week, he states, "were not intense or long continued, but came on several times during the day and night, but there were no convulsive seizures."

A letter dated June 19, 1892, reports a week of nausea during the month of April, less severe than the preceding one. He says: "I think I have gained considerably, and am told that I am looking better than usual at this time of year."

November 15, 1893. Patient visited the office for the first time in two years. He showed: Esophoria, 0.5°; no hyperphoria; adduction, 27°; abduction, 4°. The patient has had *no epileptic seizure for over two years and only one attack of nausea during 1893.* He came to New York for the purpose of being operated on for hæmorrhoids. The fact that his abduction continues to be low leads me to think that there is still some esophoria remaining in this case. He appears to be in the best of health.

A letter from the patient dated May, 1896. says that he has had no convulsive seizure and only one attack of nausea, which occurred early in 1895. This makes a total of five years without any epileptic attack and only two attacks of nausea. During the entire period *he has been principal of a school for boys, which has necessitated almost constant use of his eyes.*

CASE XI.—Mr. B., aged 22 years; single; taxidermist; sent to me on March 27, 1891.

Family History: Both parents are living. Father has had frequent epileptic seizures since his 14th year. Maternal relatives have severe headaches. No phthisical tendencies. No other cases of epilepsy in ancestry are recalled.

History of the Case: Patient began to have epilepsy during infancy, while teething. He had no more seizures until his 8th year, when he had several fits. He then went until his 14th year without another attack. Since then he has had them quite regularly. He has taken bromides for some years, but none for the past six months. During the *past six months patient has had over one hundred severe attacks.* He averages at least one a week, and often goes only five days without an attack. He is generally totally unconscious during a fit, bites his tongue, but has no aura. He has received several minor injuries by falling. He *averages at least twenty-five attacks of petit mal a day,* in which he loses consciousness, but does not fall. Patient has a good appetite, sleeps well, and has very little headache. Urine normal.

Eye-defects: Patient showed the following eye-tests : Hypermetropia, under atropine, $+ 0.75^x$; adduction, 27°; abduction, 3° $+$; sursumduction, right, 2° $+$; left, 2° $+$; manifest esophoria, 5°; no hyperphoria. Later he disclosed considerable latent esophoria.

Treatment and Results : Full correction for his hypermetropia was given for constant wear. Graduated tenotomies were performed upon both interni within a week. Results, after second operation: Esophoria, 1°; abduction, 9°; left hyperphoria, 0.25°. Up to April 6th the attendant of this patient reports that he has had no attacks of grand mal and his attacks of petit mal are much less severe and less frequent. The patient was obliged to return home, and was instructed to report again for observation after the lapse of eight or ten weeks.

On May 1. 1891, a report from the patient states : " I have had three marked attacks since I saw you,—two on April 9th and one on April 22d. I do not think I have had as many spasms as I had previous to my coming to see you. The neighbors tell me that I look greatly improved."

May 13, 1891. Patient was found dead on the banks of a small stream with a wound in his forehead, caused by a small stone on which he had apparently fallen. His hypermetropic glasses (ordered by me for constant wear) were found in his pocket.

CASE XII.—Mrs. G., aged 30 years; married; three children. Referred to me by Dr. E. W. Hedges, of Plainfield, N. J.

Family History : Unknown. No hereditary predisposition to epilepsy, as far as discovered.

History of the Case : A letter from Dr. Hedges (January 26, 1893) says : " I have advised Mr. G. to take his wife to you in hopes that you can cure her of epilepsy. She has had attacks for three or four years, following closely upon the birth of her last child. Thinking that a lacerated cervix, with angry, everted lips, might have something to do with it, I sewed up the rent ; but she is no better, and, if anything, worse. I have been so much pleased with the results of your treatment in Mr.

B.'s case that it gives me great pleasure to send you another epileptic, with the hope that you may be equally successful with her."

Subsequent conversations with her husband elicited the fact that his wife had been kept thoroughly under the influence of bromides, and that continuous epileptic seizures were liable to occur for many hours, at intervals, in spite of the drug. These paroxysms of epilepsy would often last one or two days. The patient was at this date in the South, and the reports from her were extremely unfavorable.

On March 1, 1893, this patient was brought to my office for examination.

Eye-defects: Vision was apparently normal, but, under atropine, the following refractive errors were discovered: O. D. V., $\frac{20}{50}$ —, made $\frac{20}{20}$ by $+ 1.00^{\circ} \supset + 0.50^{\circ}$ (axis, 90°); O. S. V., $\frac{20}{70}$, made $\frac{20}{20}$ by $+ 1.25^{\circ} \supset + 0.50^{\circ}$ (axis, 90°). Examination of her muscles disclosed: Adduction, 27°; abduction, 5° +; sursumduction, right, 2° —; left, 2° —; esophoria, 2°; no hyperphoria. Within two days latent esophoria of 7° disclosed itself, patient meanwhile wearing a full correction for her refraction. Her abduction ran down to 2° and her adduction increased to 33°. Within a week graduated tenotomies were performed upon both internal recti. The effect of these operations was to increase the abduction to 8° + and to decrease the manifest esophoria to 1°. It was found that the patient would not accept her full correction for hypermetropia, and her glass was decreased one-half a diopter. The physical condition of this patient was a very bad one. Her face was covered with an extreme bromide eruption; her mental powers were very sluggish; she was extremely weak and required an attendant, and was frequently attacked with paroxysms of uncontrollable sobbing and nervous excitement.

October 30, 1893. Since the last note this patient has had an attack of scarlet fever, and her two children have also been similarly attacked. She has taken no bromides since March 12, 1893. She has *entirely regained her normal mental condition*, has *increased twenty-eight pounds in weight* since her last visit, and *has had no epileptic seizure for over two months*.

During August she was somewhat irregular in wearing her glasses, and she had twenty-five hysterical seizures, some of which seemed epileptiform in type. Since that time she has worn her glasses constantly and no hysterical attacks have occurred.

November 18, 1893. Patient reports that she has had *no hysterical or epileptic seizure for three months*. She shows: Vision, $\frac{20}{15}$ in each eye with her glasses; adduction, 27°; abduction, 7°; sursumduction, right, 2°; left, 2°; esophoria, 1°. Her husband, who accompanied her, stated that he considered her as well, and that they were soon to move to the West, on account of some new business connections. A report (January, 1894) states that she remains absolutely free from hysterical or epileptic seizures.

This case has not been seen by me since the last record, but she has been completely cured of attacks of grand mal.

The following letter from her husband is on file in my records, in reply to one of inquiry from me regarding his wife's condition:—

July 10, 1896.

DEAR DR. RANNEY: I am very glad to say that Mrs. G. has not had any more of those dreadful attacks, nor, so far as I can see, any signs of them. The last was in August, 1893. Next month will make three years.

You wish to know how her present health compares with that before the eye-treatment. I do not know how to tell you, as there is no comparison. For four or five years before I took her to you she had been having attacks or seizures at irregular intervals,—sometimes one or two a month, and at others oftener. Just before she began the eye-treatment she had them frequently and violently, in consequence of which her strength was almost gone, life was a burden to her, and we had to keep a companion with her all of the time. Since November, 1893, we have had no companion, and she has done her own house-work almost continually since that time. In other words, she is a new woman, physically and mentally.

I also wish to add that, before taking her to you, I had consulted a number of prominent physicians, North and South, and they could do nothing for her. Their medicines, apparently, did more harm than good.

You made me promise to discontinue the use of all drugs, which we did, simply treating her eyes. Since that time she has used no medicine nor called in a physician, except for scarlet fever. It gives me great pleasure to write this letter, only I feel I cannot write it strong enough.

If you can make use of me in any way I shall consider it a pleasure to serve you. If it is not necessary to use our names in this matter I should prefer that you will not. However, I shall be glad to answer any letters from any one upon this subject. Sincerely yours,

Mr. G.

In order to make the diagnosis of the character of these attacks positive I wrote to the husband in October, 1896, pro-

pounding certain questions to him in regard to them.   I publish
here his reply in full:—

October 21, 1896.

DEAR DOCTOR: I will take your questions in the order in which you have
asked them, and try to give you replies to each :—

Did Mrs. G. ever give a cry as her attacks came on ?   No.

Did she ever become completely unconscious?   Yes ; always.

Did she ever bite her tongue?   Yes ; several times.

Did she ever froth at the mouth?   Yes ; but not very much.

Was there any blood in the froth?   Yes ; I suppose from biting her
tongue.

Was there contraction of muscles of arms, legs, and face?   Yes ; always
in each ; and her mouth was always drawn very much to one side.

Did she ever fall?   Yes, several times ; and would have fallen every time,
unless she was caught or was sitting or lying down.

Was there ever lividity of the face ?   Yes.

Did she ever have any warning of the attacks?   This is one thing I could
never find out.   She has always had such a dread of the attacks that she will not
talk of them.   Many times I have asked her this question, and she would say
"No"; but I have noticed that, on the days that she would have the seizures,
she would be different than at other times ; that she would have a frightened,
nervous look that would lead me to suppose that she had some feelings that made
her fear an attack.   This is all the warning that I have ever known of.

She commenced having the attacks in February, 1889.   They continued
until you took her in hand in February or March, 1893, coming at intervals of a
few weeks apart all of this time, except when under the strong influence of
bromide.   Yours, with sincere respect,

MR. G.

A letter from Dr. Hedges regarding this patient, dated
July 7, 1896. says: "Mrs. G. moved away about two years ago,
since which time I have heard nothing as to her condition.
Just previous to leaving she called upon me and declared that
she was a different woman, mentally and physically, since the
operation ; that it seemed to her as though she had been living
in a dream for years past.   She had gained about thirty pounds
in weight. as I remember it, and certainly looked younger and
better than I had ever seen her.   At that time there had been
no return whatever of the epilepsy."

CASE XIII.—Mrs. W., aged 30 years; married; three children.   Referred to me. on April 8, 1893, by Prof. A. A. Smith,
M.D., of New York City.

Family History : Mother had considerable headache; father

is strong and healthy. No hereditary tendency toward nervous
diseases or phthisis.

History of the Case: Urine normal. After the birth of
her last child this patient began to have epileptic seizures while
residing in high altitudes in the West. She was placed under
bromides, but the seizures continued at varying intervals and
tended to increase rather than diminish. She came to New
York for medicinal treatment, and was referred to me later to
see if any light could be thrown on the exciting causes of her
attacks. While no very satisfactory record of the frequency of
her attacks could be obtained, the reports made by her husband
during the first week that she was under my personal observation show that several marked epileptic seizures occurred within
the first week. Her husband also stated that he was afraid, on
account of the frequency of her epileptic seizures, to have her
take her meals in the public dining-room of a hotel, as he never
felt sure when they would not occur. Generally the seizures
had occurred immediately after awakening from a night's sleep.

Eye-defects: On the first visit to my office (April 8, 1893)
this patient showed emmetropia, having $\frac{20}{15}$ vision in each eye,
under homatropine. and tolerating no plus glass. Adduction,
23°; abduction, 6°; sursumduction, right, 2° — ; left, 2° — ;
no esophoria; no hyperphoria. Although the patient showed
for two days no manifest muscular defect, her low abduction led
me to suspect a latent esophoria. The patient was, therefore,
given a 1° esophorial prism for twenty-four hours. The following day she disclosed: Esophoria, 1.5°; abduction, 4° +. Within
ten days (by the judicious wearing of prisms) she showed, over
a 5° prism: Esophoria, 7°; no hyperphoria; adduction, 39°;
abduction, 0; and homonymous diplopia whenever a red glass
was placed before either eye without the prism.

I was myself somewhat surprised to find, at the beginning
of the investigation of this patient's eyes, an apparent orthophoria with an absolute emmetropia. The serious malady with
which she was afflicted and its persistency (in spite of medication) led me naturally to hope for a marked refractive error, and
probably, also, a manifest heterophoria. It took fully ten days
of careful observation, in this case, before I felt that I was
justified in performing a graduated tenotomy or even in extend-
ing to the patient any basis of hope for improvement from the
eye-treatment.

Treatment and Results: On April 19, 1893, I performed a graduated tenotomy upon the right internal rectus. Result, after operation: Esophoria, 1°; no hyperphoria; abduction, 10°—. Three days later this patient showed: Esophoria, 0.5°; no hyperphoria; adduction, 33°; abduction, 8°; sursumduction, right, 2°; left, 2°. Since this operation *no epileptic seizures have occurred.* The patient was seen, at intervals, during the subsequent nine months, and very little change in her eye-tests has been observed. At her last visit to me, at that time, she stated that she was, in her opinion, "absolutely well."

Since her return to the West this patient has been entirely cured (up to the last report in January, 1897) of genuine epilepsy by the relief of eye-strain through one graduated tenotomy.

Several points of great interest are illustrated by this case. In the first place, she had no error of focus, either prior to or after the instillation of atropine. In other words, she was absolutely emmetropic. No carping critic can, therefore, lay any stress upon the exact amount of benefit that must be attributed to the glasses prescribed. In the second place, the esophoria was almost totally latent. Seven degrees were disclosed later by the patient, after repeated examinations. She would, therefore, have been pronounced by many oculists as free from any muscular trouble. In the third place, the eyes were brought to a state of perfect muscular adjustment by one operation made upon the the right internal rectus. In the fourth place, no further latent muscular defect in the orbit has ever been observed. In the fifth place, the fits ceased at once after the tenotomy, and have never returned, although four years have elapsed. In the sixth place, her physical condition has been made perfect and remained so. She had been an invalid. Finally, she had "genuine" epilepsy, as all my critics must allow upon the evidence here presented.

I quote first a letter from the husband, in answer to my written inquiry concerning the patient, as follows:—

July 11, 1896.

MY DEAR DR. RANNEY: Your kind letter of the 7th inst. came yesterday. Mrs. W. was so much pleased that she said she would answer it herself. She has never felt better or looked better than she does at present. The only symptom she has had was in July, 1895, on board of a steamer, the particulars of which I wrote you shortly afterward. The food had been wretched, and for nearly ten days she had got along by making tea in her cabin,—lunches, in fact. She could

not swallow the "hash" prepared on the boat; so that one morning, while dressing, she fell over, till I caught her, helped her on the bed, and she simply kept
quiet the rest of that day.

[NOTE.—This, in Dr. Smith's opinion and my own, was only a slight faint.
It was not convulsive, nor was any loss of consciousness noted.]

Yours sincerely,

MR. W.

In response to a letter to Prof. A. A. Smith, M.D., asking
him to write me concerning the attacks of this patient, I received
a personal call from him. He stated at that interview that, unfortunately, he had not actually seen the patient in any of her
epileptic seizures, but had sent her to me for eye-treatment, because he believed them to be attacks of genuine epilepsy, and
had told the family that drugs offered no prospect of any
permanent benefit, in his opinion. He advised me to establish
the actual type of convulsion that the patient had, by propound-
ing to the husband by letter certain questions that he suggested.
I did so, and these were the questions and answers:—

1. In her attacks did she lose consciousness completely?
" Yes."

2. Did she ever make any noise or give any cry as the
attack came on? " Yes."

3. Did she froth at the mouth during the attacks? "Yes,
a little."

4. Was there ever a tinge of blood in the froth or on the
pillow? "Yes, but not always."

5. Was there any soreness of the tongue after the attack?
"I cannot say."

6. Was there much convulsive movement of arms and
legs? " Yes."

This last report was followed by an unexpected visit from
the patient herself. She was the picture of health, and has had
no symptoms of her old malady.

CASE XIV.—Mr. S., aged 19 years; single; telegraph
operator. Referred to me on March 2, 1893, by Dr. J. B. Hulett,
of Middletown, N. Y.

Family History: Father died of Bright's disease. Mother
has some headache. No hereditary tendencies to nervous diseases so far as known.

History of the Case: Urine normal. Up to the 10th year of age the patient was perfectly well. At that time he had what was supposed to be a fainting-spell. At 12 years of age the patient had his first severe epileptic seizure. Another followed in a few months, and he was put under bromides. For nearly seven years he took the bromides constantly, but continued to have epileptic seizures at varying intervals. Occasionally he would go three months without a seizure, but more often he would have one every two weeks. For the past three months he has taken no bromides, and the frequency of his attacks has not been materially altered by their withdrawal. Two weeks before coming to me he had two severe attacks in one day.

Eye-defects: This patient had $\frac{20}{15}$ vision in each eye under homatropine and tolerated no plus glass. He showed on the first examination : Left hyperphoria, 0.25° ; esophoria, 0.25° ; adduction, 25° ; abduction, 6° ; sursumduction, right, 2° ; left, 2°. On account of his low abduction, latent esophoria was suspected. Under the influence of a 1° esophorial prism, which was worn by the patient for twenty minutes, he showed : Esophoria, 3.5° ; no hyperphoria ; abduction, 5°. Within a few days he wore with comfort a 5° esophorial prism and showed : Esophoria, 7° ; adduction, 37°; abduction, 2° ; no hyperphoria.

Treatment and Results: On March 17, 1893, a graduated tenotomy was performed on the right internal rectus. After the operation the patient showed : No esophoria ; abduction, 9°. Two days later the patient showed : Esophoria, 0.5° ; adduction, 37° ; abduction, 8°. The patient went home and resumed his vocation, with instructions to return after the lapse of a few weeks.

April 17, 1893. Patient visited my office and reported that he *has had no attacks during the past five weeks.* He shows : Esophoria, 0.5° ; no hyperphoria ; adduction, 38° ; abduction, 6° + ; sursumduction, right, 2° ; left, 2°.

On April 25, 1893, this patient was seized with an epileptic attack. During the fit he overturned a lamp, which set fire to his clothing and burned him so that his life for a time was despaired of. On December 2, 1893, this patient reported at my office for further examination of his eye-defects. He states that he *passed over six months without any epileptic seizure.* During October, 1893, a slight seizure appeared, but since then he has been apparently well and does not even have any of his old apprehensions of an approaching epileptic fit. He discloses no hyperphoria; esophoria, 1°; adduction, 30°; abduction, 7°.

On January 9, 1894, the patient reported that he had passed three months with no attack, and only one epileptic seizure had occurred in over nine months.

This patient was having an attack of typical grand mal on an average of every fourteen days when I first saw him. He had taken bromides for seven years. His eyes were perfectly normal in construction. Even under atropine he showed emmetropia. He had a high degree of latent esophoria. If "counter-irritation" is the reason why graduated tenotomies help epileptics (as some critics would lead others to believe) this patient had a full dose. One hand was nearly burned off by overturning a lamp while in a fit, amputations of fingers were required, and his life was in peril for some time. His condition to-day warrants (in my opinion) the report of "decided amelioration" by eye-treatment (see table). He has passed at one time over six months without a seizure, and has had about one-fourth of the attacks during the past three years and a half that his previous average of attacks would have aggregated in the same time.

CASE XV.—Mr. P., aged 26 years; single; merchant.

Family History: Maternal relatives are all more or less nervous and most of them have eye-defects. Father has very marked difference in refraction of the two eyes with an approach to vertical strabismus. One brother has had a few attacks of epilepsy. One sister was a chronic invalid for many years, until her eye-defects were rectified by me.

History of the Case: This patient's first attack occurred in 1883, and he has had them at irregular intervals ever since. They occur always at night. He usually bites his tongue during the attack. This is often the only way that he knows he has had an attack, as he finds a little blood on the pillow. *During the past year he has had six severe epileptic seizures.* He has taken bromides during the past few years without any apparent influence upon his attacks.

He has been troubled all his life with the most obstinate form of chronic constipation, for which he has taken cathartics regularly every night at the advice of different physicians.

This finally produced piles so badly that several operations were necessary for their removal. He has always noticed that he is more liable to have an attack when he is troubled the most with constipation.

He has consulted several prominent neurologists, all of whom have told him that his attacks were genuine epilepsy and incurable except by bromides. His mother finally became discouraged with the negative results of the bromide treatment and brought him to me for diagnosis and treatment on September 7, 1892.

Eye-defects: At the first visit this patient showed the following eye-tests: Vision under homatropine $\frac{20}{20}$ in each eye; slightly improved by + 0.50ᶜ (axis, 90°); no hyperphoria; esophoria, 2°; adduction, 23°; abduction, 5°; sursumduction, right, 2° + ; left, 2° +. Within a week, while wearing a 5° esophorial prism, he showed: Esophoria, 7°; no abduction; adduction, 35°. *Unconquerable homonymous diplopia* when the prisms were removed. Later he disclosed a very high degree of latent esophoria.

Treatment and Results : Three graduated tenotomies were performed,—two on the right and one on the left internal rectus. The latent esophoria was developed by the judicious use of prisms before each operation. The last operation was performed on November 18, 1892.

Since that time the patient has been seen at frequent intervals and his eye-muscles carefully tested; but he shows a condition of practical orthophoria. His tests at his last visit, September 14, 1893, were: No esophoria; no hyperphoria; adduction, 45°; abduction, 8°— ; sursumduction, right, 2° + ; left, 2° +.

The patient had an epileptic seizure the day after the first examination of his eyes, and another light attack in January, 1893. *For months he had no epileptic attack* of any kind and reported that he had a movement of the bowels every day regularly without cathartics, had abandoned all drugs for that purpose, had gained very markedly in weight, and considered himself perfectly well. He has taken no bromide salts since his first visit to my office. On December 1, 1893, this patient had an epileptic seizure that was quite severe, followed by vomiting. I am informed that he had been exercising violently each day for some months and had lately been taking codliver-oil without my knowledge. That seemed to cause a gastric derangement.

This patient has been practically cured of epilepsy. He is rather hard to control, and does not follow instructions as to regularity of habit, eating, sleeping, and exercise. He keeps very late hours at times, is entered often in trial contests of skill of a violent athletic kind, eats irregularly and too heartily of rich foods, and in many other ways brings about an occasional gastric upset and a very rare epileptic seizure by his own acts. If he lived a regular life, I believe he would never have an epileptic seizure. Furthermore, he uses his eyes constantly, as a book-keeper, during business-hours.

CASE XVI.—Mr. II., aged 18 years; referred to me by Dr. C. E. Stephens, of Syracuse. N. Y., on March 23, 1893.

Family History: No hereditary tendencies to nervous diseases, so far as is known.

History of the Case: Patient has been afflicted with attacks of petit mal for a number of years, and for the past few years has had numerous attacks of grand mal. Bromides were tried, but had no marked influence upon the attacks. When he came to me for examination he was averaging one attack of grand mal a month, and from two to fifteen attacks of petit mal a day. These attacks of petit mal were described by him as "whirling attacks," in which objects appeared to dance before his eyes, whether open or shut, in various fantastic forms. A number of these seizures were observed in the office and found to be genuine attacks of petit mal. His conversation would be abruptly stopped; his head would turn slightly to the left with some twitching of the mouth; he would mutter a few incoherent words, rub his eyes, and in a minute or two return to his former condition of mind. He is very sluggish mentally, but perfectly rational. The whole aspect of the patient was that of a person who was somewhat defective mentally. His movements were slow and he had a peculiar drawling articulation.

Eye-defects: The eye-conditions of this patient were particularly difficult to solve. It seemed almost impossible to get any satisfactory tests either for refraction or heterophoria. Patient had marked divergent strabismus, and before atropine was used his vision was only $\frac{20}{40}$ — and could not be improved.

Under full atropine his vision was made $\frac{20}{50}$ — by + 2.50° in the right eye and + 1.00° in the left eye. No accurate muscular tests could be obtained, as the patient could not be made to see two images even with a red glass. He was very slow in speech and had several " whirling " attacks during the tests. An ophthalmoscopic examination showed no abnormal condition of the retina.

Treatment and Results: A full correction was given for his hypermetropia, and atropine used until he would tolerate it. Free graduated tenotomies were performed upon both external recti. After the operations patient showed very little lateral jump on exclusion, but apparently some vertical jump for right hyperphoria. He was sent home with instructions to keep up the use of atropine and report weekly.

October 25, 1893. Up to this date the patient has had but two attacks of grand mal since March last, having gone at one time four months without an attack. The patient seems to think that his attacks of petit mal. or " whirling attacks," have been much less frequent. It is still impossible to obtain any accurate muscular tests. as the patient does not seem to have binocular vision. He uses either eye at will, but seems unable to see with both eyes at once. His face indicates a right hyperphoria and he has a slight vertical jump on exclusion. A 2° right hyperphorial prism was combined with his distance glasses and the patient sent home for a few months. The inability to get any muscular tests makes this case a very difficult one and leaves me in doubt as to what his future treatment will be.

This case, in justice to myself. should be excluded from the list of reflex epilepsies, as he is one of organic brain disease. Furthermore, he was under treatment but a short time, and no satisfactory eye-tests were ever obtained, as his mental powers were too much impaired to make his tests reliable. He is reported by me as " unimproved."

CASE XVII.—Miss F., aged 17 years. Referred to me on December 28. 1891.

Family History: Mother has hypermetropia, astigmatism, and esophoria. No hereditary tendency to nervous diseases so far as known.

History of the Case: When 7 years old the patient had three convulsions within six months. From that time she had no attacks until February, 1891, when she had another. Since then she has had one in April, one in September, and one in December. These attacks occur at the menstrual period and are always accompanied by nausea. She has been examined by several physicians and no organic disease found.

Eye-defects: Patient had $\frac{20}{15}$ — vision under atropine and would tolerate no + glass. Her muscular tests on the first visit to the office were: Adduction. 25°; abduction, 8°—; sursumduction, right, 2°; left, 2°; no hyperphoria; esophoria, 3°. Within the next three weeks, after wearing prisms for the manifest esophoria (which were slowly increased in strength), she showed, with a 4° esophorial prism: Esophoria, 5°; no hyperphoria; adduction, 33°; abduction, 5°—. In accommodation: Esophoria, 10°.

Treatment and Results: A graduated tenotomy was performed on the right internal rectus with an immediate result of: Exophoria, 0.25°; abduction, 10°. Six days later she showed: Adduction, 32°; abduction. 8°; no hyperphoria; no exophoria. During the next three months the patient was seen very frequently, and prisms were tried for a suspected latent esophoria, but no esophoria was discovered. Her tests on May 3, 1892, three months after the operation. were: Adduction, 33°; abduction, 8°; no hyperphoria; esophoria, 0.5°; in accommodation, esophoria, 4°. The results of treatment in this case were negative. She had one attack in February and another in July. At her last report she was having an attack about once in five months.

This patient is also reported in the table as unimproved, although the opportunities for accurate eye-work were in no way hampered for one year. I am satisfied that some other reflex causes existed in this case (possibly pelvic). Her epileptic seizures were too infrequent to enable me to tell what clinical results I was to expect from each tenotomy. She was absolutely emmetropic, but had a very high degree of latent esophoria. This was relieved satisfactorily, yet her fits continued.

CASE XVIII.—Miss J., aged 22 years. Referred to me on October 13, 1890, by Dr. H. J. Dwinell, of Barton, Vermont.

Family History: Mother has a high degree of hypermetropia. Father has marked exophoria. Paternal grandfather had ptosis. One paternal aunt has hysterical attacks resembling epilepsy. Maternal grandmother died of phthisis.

History of the Case: When the patient was a baby she had an eye that turned down at times so that the pupil could not be seen. When 8 years old she began to have slight "fainting attacks," which steadily grew more frequent and of longer duration. About five years ago she began to have convulsions with contraction of arms, legs, and face, her head drawing to the right, with drooling at the mouth. She was at once put upon bromides and only averaged four severe attacks a year. At this time she is not taking bromide and averages one attack a month.

Patient has a severe headache once a month and slight headaches between. She has always been a very nervous child and sleeps very poorly. She complains of a weak back, pain in her left shoulder, and a tickling in her feet and legs which is very annoying. Patient is very moody; is unable to do even light work about the house, and never goes out without an attendant.

Eye-defects: At the first visit patient showed a *crossed diplopia* of 12° and a *vertical diplopia* of 2° with a red glass; vision, $\frac{20}{30}$ in each eye. Under atropine vision was $\frac{20}{30}$ with $+ 1.00$. Within two days she disclosed: Right hyperphoria, 5° $+$; exophoria, 18°; abduction, 20°. The other muscular tests could not be obtained on account of the extreme heterophoria.

Treatment and Results: The treatment of this case consisted solely in graduated tenotomies for the correction of the heterophoria. An operation was performed upon the right superior rectus, the following day upon the left external rectus, and six days later upon the right external rectus.

Three days after the last operation patient showed: Adduction, 25°; abduction, 5° —; sursumduction, right, 3°; left, 3°; no hyperphoria; exophoria, 3°. Patient was then sent home with instructions to report monthly and to return in the spring to complete the treatment.

January 12, 1891. Patient reports only two attacks since October 17, 1890. These were associated with menstruation. She has gained markedly in weight. The last attack occurred on December 12, 1890.

March 6, 1891. Patient's father states that she has had no convulsion since December 12, 1890,—nearly three months.

She has momentary flashes about a week before menstruation. She is much less moody, goes where she chooses without an attendant, and visits places of amusement alone and friends for a day at a time.

April 18, 1891. A letter from her father says: "Our daughter has had no return of the disease since last note,—not even a symptom or indication of the return of the disease since the last week in February. She sleeps without that movement of the muscles. The 'disagreeable smell' which troubled her so much does not return, and she is free from aches and pains. No headache. All this without one drop of medicine since the 1st of October last."

September 24, 1891. Patient returned to-day for further treatment. She reports that only five severe attacks have occurred since last October, the final one being on August 25th. She has had a number of attacks of petit mal since last attack of grand mal. Patient shows: Adduction, 23°; abduction, 6°; sursumduction, right, 2°; left, 2°; right hyperphoria, 0.25°; exophoria, 2°.

A full correction of her hypermetropia ($+ 1.00^s$) was given for constant wear and her muscles exercised with prisms. In five days she showed: Adduction, 28°; abduction, 6°; sursumduction, right, 2°; left, 2°; no hyperphoria; no exophoria.

November 23, 1891. A letter from her father says: "Our daughter is doing finely; only one attack or convulsion since last August. Her general health is much better than when in New York. She works every day and has commenced to practice on the piano. We can see a great improvement from what she was before you saw her the first time."

March 12, 1893. The father states, in a letter: "My daughter is much improved since our visit to you in the fall of 1890, and, *since October, 1891, she has had but three convulsions*, and these not as severe as those she had before you saw her."

I have quoted several extracts from some of his letters in my reports of this case up to February, 1894. She went at one time eighteen months with only three light attacks; had resumed practice on the piano; required no attendant, as she once did; went to places of amusement, as did her friends; and was actively employed in house-work during the day. Her general health had been almost completely restored.

During 1894 this patient had a number of convulsions, and in March, 1895, was again brought to New York. She disclosed again a large amount of exophoria, and another tenotomy was performed upon the right external rectus. This did not entirely relieve the exophoria, and a 2° exophorial prism was conbined with the spherical glass in each eye. She was then sent home with instructions to return in the fall for further treatment. As she continued to have an occasional convulsion after the last operation, her parents lost their hope of a cure and abandoned treatment. Since that date no report has been received from her and I suppose she continues to have an occasional attack. This case gave promise of being a most favorable one, if treatment had been continued until all heterophoria was removed; and it was with great regret that I saw the treatment stopped when so near completion. The final results of eye-treatment in this case I do not regard as yet established, but enough benefit has already been gained to make an impression upon every one who had seen her often in the past.

It should be remembered by the reader that, when this patient came to me, she showed both crossed and vertical diplopia, accompanied by severe attacks of grand mal and marked evidences of the poisonous effects of the bromide treatment. Her father came at the same time (with almost identical eye-tests) as a terrible sufferer from headaches. He was completely cured by tenotomies and has remained well up to our last records of his case. The treatment of this case was never completed to my satisfaction. It was a very difficult eye-problem to handle, and long intervals of rest between each operative step were deemed by me to be the safest way to establish a perfect adjustment. The father seemed to fail to appreciate the time and skill required to do this work, and ceased to follow up the eye-treatment (after an exhibition of pique on his part).

CASE XIX.—Miss D., aged 12 years. Referred to me by Dr. T. J. Martin, of Buffalo. N. Y., on March 15, 1893.

Family History : No hereditary tendencies to nervous disease, so far as known.

History of the Case: This patient had her first attack while teething, her second when 3 years old, her third when 6 years old, and her fourth when 10 years old. In 1891 she began to have attacks every six weeks and was put on bromides. She then went eighteen months without an attack, and bromides were stopped, as the attending physician deemed it wise to discontinue them.

During the three months prior to her first visit to me she had four severe attacks of typical epilepsy. She is otherwise a strong, well-developed, healthy child.

Eye-defects : Under atropine patient showed : O. D., + 2.75^s ; O. S., + 3.25^s ; adduction, 25° ; abduction, 6° ; sursumduction, right, 2° ; left, 2° ; no hyperphoria ; esophoria, 2°. Within a few hours, under the influence of esophorial prisms, she showed : Esophoria, 7.5° ; abduction, 2°.

Treatment and Results : The following glasses were given for constant wear, with instructions to use atropine if ciliary spasm should occur : O. D.. + 2.00^s ; O. S.. + 2.50^s. Graduated tenotomies were performed upon both interni within a week and the patient sent home for a few weeks.

May 5th. Six weeks after the operation the patient returned for further treatment and her tests were : Adduction, 26° ; abduction, 8° ; sursumduction, right, 2° ; left, 2° ; no hyperphoria ; esophoria, 0.25°. Her father reports that she has had but one attack (April 25th), which was lighter than usual. Nothing was done at this visit except to exercise her muscles with prisms.

November 27th. Patient's mother reports that she has had five attacks since last April (a period of over seven months) ; that the attacks have been lighter, and the interval between attacks much longer, than prior to the eye-treatment.

As she seemed to tolerate a stronger spherical glass, she was given O. D., + 2.50^s ; O. S., + 3.00^s ; for constant wear. Her muscular tests were : Adduction, 25° ; abduction, 8° ; sursumduction, right, 3° ; left, 3° ; no hyperphoria ; no exophoria.

For about two years I have not seen this child, and the eye-treatment has been only partially carried out. She has passed some quite long periods without attacks since she was first seen by me, but of late has had epileptic seizures more frequently than before. Her physician and parents attribute their increase to the approach of menstruation and to overloading her stomach. I have reason to believe that a completion of the eye-work would lead to still better results than have thus far been obtained.

CASE XX.—Miss D., aged 30 ; bank-clerk. Referred to me on June 15, 1893, by Dr. Clara E. Gary, of Boston, Mass.

Family History: Mother has had violent sick headaches. Father has been a victim to "blind" headaches and eye-trouble. Headaches are very common in maternal ancestry. One brother and two sisters died at 2 years of age,—one from "water on the brain," one from meningitis, and one in convulsions.

History of the Case: From two to three years of age the patient "has always had something the matter with her stomach." When 13 years old she began to have severe sick headaches, and these have continued ever since. Menstruation began at 18 years of age. Two years later, during menstruation, she had her first epileptic seizure, and during the next five years she had four severe attacks.

Five years ago she had two attacks in one day, and was put on bromides, which she continued to take up to October, 1892. During these five years she had ten severe attacks, the interval between attacks varying from two months to a year.

Since the stoppage of bromides she has had two attacks,— one in March and the last one in April. All of her attacks are accompanied by a total loss of consciousness, rigidity, frothing of mouth, slight contraction of muscles, and no aura.

Eye-defects: Under homatropine the patient showed: Hypermetropia, + 1.50° ; adduction, 37° ; abduction, 5° ; sursumduction, right, 3° ; left, 3° ; no hyperphoria ; esophoria, 9°. After wearing a 4° esophorial prism for twenty-four hours her esophoria was 12°, her abduction dropped to 3°, and she had homonymous diplopia with the red glass.

Treatment and Results: As the patient was able to remain in New York but three days, she was given + 1.00° ◯ 2° prism, base out, in each eye for constant wear, and was instructed to return as soon as possible for operation.

October 9, 1893. Patient returned to-day for further treatment. She shows esophoria of 11° and abduction of 2° —. Within a few days graduated tenotomies were performed upon both internal recti, and the prisms were removed from her glasses. Directly after the last operation she showed: Esophoria, 2° ; abduction, 8°. One week later her tests were: Adduction, 38° ; abduction, 6° ; no hyperphoria ; esophoria, 2.5°.

November 14, 1893. Patient writes that she has had no

epileptic seizures since last April (a period of over seven months), and has been doing her work at the bank regularly.

This patient has been completely cured of her epileptic seizures.

She had "genuine" attacks of grand mal, with frothing at mouth, total loss of consciousness, and rigidity, followed by clonic spasms of arms and legs. She did not always bite her tongue, and she had no special aura. She had taken various combinations of bromides for five years. In spite of large doses, she had ten severe convulsions during that period; and after stopping the bromides she had two severe fits within the space of two months. From the first visit to my office up to the present time she has not had a single convulsion or any symptoms of one. In answer to a letter of inquiry from me she writes as follows:—

July 12, 1896.

DEAR DR. RANNEY : If all had not been well with me, I think you would have heard. I am very glad to be able to say that I have never had a return of the old trouble, and only one sick headache in all the time since you operated upon my eyes.

It is such a comfort to be free from these lesser ills, and, for the greater one that hung over me for so long a time, no one can tell what a feeling of thankfulness there is, except one who has had the trouble and been freed.

I have been to the physician who first had charge of my case to tell him of the success of your treatment, hoping that others might be helped who came under his care.

The time has been so long now I feel that I may call it a cure, although I hardly dare do so. Am I right?

Very truly yours,
Miss D.

No letter has thus far been received from Dr. Gary in reply to one sent to her former address, but the letter from the patient tells its own remarkable story in a very simple and direct way. This patient has used her eyes constantly as a book-keeper.

CASE XXI.—Mr. T., aged 22 years; married; photographer. Referred to me on February 17, 1892, by Dr. Elmer Small, of Belfast, Me.

Family History : The father has severe sick headaches and "dizzy spells." Mother and two brothers are perfectly well. One paternal aunt has severe sick headaches. One paternal uncle has nervous prostration.

History of the Case: Patient was a very strong child, but had momentary attacks of unconsciousness. When 12 years old he had his first typical epileptic seizure, with the epileptic cry, loss of consciousness, violent convulsive movements, and frothing at the mouth. He was at once put under bromides, and they have been kept up ever since. He averages about six severe seizures a year; the shortest interval between attacks has been two weeks and the longest nine months. During the past year he has also had several slight attacks of unconsciousness, lasting only a few seconds. Patient has always had some difficulty in using his eyes, and three years ago consulted Dr. E. E. Holt, of Portland, Maine. Dr. Holt prescribed myopic cylinders, and cut both internal recti.

Eye-defects: Under atropine patient showed: $+ 0.50 \supset +$ 0.50° (axis, 90°) in each eye; adduction, 40°; abduction, 6°; sursumduction, right, 2° —; left, 2° —; no hyperphoria; esophoria, 2°. Within a week, by the use of esophorial prisms, the patient showed, while wearing 5° of prism: Esophoria, 7°; adduction, 60°; abduction, 1°.

Treatment and Results: Patient was given hypermetropic cylinders ($+ 0.50^c$; axis, 90°) for constant wear. Later on this was increased to a full atropine correction. A graduated tenotomy was performed on the left internal rectus, and five days after the operation he showed: Adduction, 56°; abduction, 8° —; no hyperphoria; esophoria, 1°. Since the last operation the patient has been seen at frequent intervals, and prisms have been given several times for suspected latent esophoria; but he has never manifested clearly any esophoria, and the prisms have been removed. His last test on September 25, 1893, showed: Adduction, 50°; abduction, 7° +; no hyperphoria; esophoria, 0.5°. All bromides were stopped from the time of his first visit (February, 1892). During the first year he had fourteen attacks; but they have been steadily decreasing in number, so that for the past year he has had only six attacks. This is the exact number that he had when under heavy doses of the bromide salts. The physical condition of this patient has improved greatly since the bromides have been withdrawn.

In actual number I do not think the attacks have been very markedly decreased since the last note (1894), although the physical condition of the patient had improved to a remarkable degree when I last saw him. Suspecting that an old injury to his head might be a cause of his epileptic seizures. I sent him to Dr. Robert F. Weir with the following note:—

March 13, 1895.

My dear Dr. Weir : The bearer, Mr. T., is an epileptic. The rectification of his eye-muscles has done him much good, but has not arrested his seizures *in toto*.

He gives a curious history of a fall upon his head prior to the epileptic attacks, and has come from Maine to see if trephining would help him.

I would appreciate a written opinion from you as to what you would advise ; and, should you choose to act, I would intrust him to your care.

Cordially yours,

A. L. RANNEY.

To this letter Dr. Weir sent me the following reply :—

DEAR DOCTOR : This patient cannot describe his fits sufficiently in detail to attempt any localization, and the point of supposed trauma is too indefinite for action. He should, in my judgment, be seen in several fits and the sequence of muscular invasion noted down. I may, however, say that, unless thus positively localized, trephining will do but little good ; and even when localized and treated surgically the improvement is a dubious one.

Yours most truly,

R. F. WEIR.

CASE XXII.—Miss R., aged 22 years. Referred to me on October 24, 1893.

Family History : Mother and two maternal aunts died of phthisis. Father has chorea and has had tenotomies performed on one internal rectus for esophoria. Chorea for at least four generations on the paternal side. Paternal grandmother and one great uncle committed suicide. Several cases of insanity among the paternal ancestors.

History of the Case : When the patient was only a day old she was given an overdose of morphine. This caused convulsions, which lasted for several days. This child again had one convulsion during her second summer while teething, and not long afterward her eyes became markedly crossed. When 5 or 6 weeks old tenotomies were performed upon both interni by the old-fashioned method, for the correction of this defect. Although nervous and irritable, she grew large and apparently well, developing early. In her thirteenth year she began to have "attacks of sickness at the stomach, or perhaps sour

risings." These continued to increase in frequency, and four
years ago she was placed in care of a physician, who pronounced them attacks of petit mal. During the past four years
the spasms have changed in character and become extremely
frequent. The patient is very regular in her menstruation, but
suffers from very annoying chronic constipation.

Eye-defects: At the first visit patient showed : Vision, O.
D., $\frac{20}{50}$ — ; O. S., $\frac{20}{30}$ — , not improved by any glass ; left hyperphoria, 5° ; esophoria, 7° to 15° ; extreme nystagmus when either
eye is covered, but only noticeable at times when using both
eyes. Her left eye rolled up very markedly when covered, and
putting a red glass or the Maddox rod before either eye caused
such marked nystagmus that those tests were impossible. I was
also unable to determine the strength of her ocular muscles.

A 4° left hyperphorial prism was given and in thirty minutes she showed left hyperphoria of 9°. Under homatropine
her vision in either eye was made $\frac{20}{30}$ — by + 1.25ˢ. Binocular
vision, $\frac{20}{30}$ — .

Treatment and Results : The treatment consisted in the
correction of her hypermetropia by glasses (+ 1.00ˢ) to be
worn constantly, and a graduated tenotomy upon the left superior rectus. Immediately after the operation she showed 3° of
left hyperphoria. Ten days after the operation she showed :
Left hyperphoria, 1° ; esophoria, 10° to 14° ; and the eye did
not roll up when covered. She was then sent home with instructions to report monthly and return later for treatment.

November 13, 1893. A letter from her father states that
she has had six attacks of petit mal during the past sixteen
days. This is somewhat less than prior to the tenotomy.

The treatment of this case can hardly be considered as
more than begun. The heterophoria is so extreme that, until
further work is done upon the eye-muscles, the nystagmus and
the nervous symptoms will undoubtedly persist.

The extreme difficulties in the treatment of such a case,
when the eyes had already been operated upon years ago and
after nystagmus had set in, must be recognized by all who have
had any experience in the treatment of eye-muscles. I saw this
case last week. She has fewer petit-mal attacks each month
than when my work was commenced upon her eyes, and an
almost total arrest of the nystagmus so long as the right eye is
not covered. She shows still some tendency toward esophoria

and left hyperphoria. She is now wearing prisms because I have felt a hesitancy in doing any tenotomies upon this patient without long intervals of rest between each step. I have reported this case as one of "decided amelioration" in the table that follows.

CASE XXIII.—Mr. M., aged 31 years; single; clergyman. Referred to me on October 21, 1893, by Dr. A. A. Smith, of New York.

Family History: Mother had severe neuralgia. Father and three brothers are all living and well. Maternal aunt had epilepsy. One case of insanity in ancestry.

History of the Case: Patient was well and strong up to 18 years of age, when he had a "fainting attack." At 24, while in college, he had his first regular epileptic seizure, during a class-rush, after a hard night's work. A few weeks later he had a second attack in presence of a doctor on a train. He fell with the epileptic cry, had marked convulsive movements, frothed at the mouth, was unconscious for twenty minutes, and very sleepy for hours afterward. At that time he was put on heavy doses of bromide and has kept it up pretty constantly ever since.

Three years ago he had attacks from one to three weeks apart, but last year had only four severe attacks. He always has his attacks in the morning and very often on the cars. Looking out of a car-window always causes great discomfort.

Eye-defects: Under homatropine the patient has $\frac{20}{30}$ vision, but tolerates a + 0.50°. His muscular tests at the first visit were: Adduction, 23°; abduction, 4°; sursumduction, right, 2° —; left, 2° —; right hyperphoria, a shade; esophoria, 1°. Within a week, after wearing esophorial prisms, he showed, while wearing 6° of prism: Adduction, 35°; no abduction; no hyperphoria; esophoria, 7.5°. *Unconquerable homonymous diplopia* existed when the prisms were removed. Later a very high degree of latent esophoria disclosed itself.

Treatment and Results: The treatment of this case consisted of three graduated tenotomies on the internal recti muscles. One on each internus was performed within a week; and two weeks later the right internus was again button-holed. Between these steps the patient was seen daily; his eye-muscles were exercised with prisms, and prisms were given as fast as he showed any esophoria.

Five days after the last operation he showed: Adduction, 41°; abduction, 6° +; no hyperphoria; no esophoria. All bromides were stopped on November 8th.

December 2, 1893. Patient reports that he has had no epileptic attack. His eye-tests are the same as last noted. He was sent home with instructions to render a monthly report for three or four months. There was reason to believe that some latent esophoria still existed, but I deemed it wise to allow the patient ample time to learn to use his eye-muscles in their modified adjustment before I investigated the suspected, but entirely latent, heterophoria.

On January 3, 1894, a report was received from this patient. He says: " I have been entirely free from any epileptic seizures, and I find my head much relieved when at work. I can speak and write with a facility I have not known for years."

This young minister of the gospel had typical attacks of genuine epilepsy, and had been obliged to cease his work as a pastor. While in New York City to get professional opinions regarding his case he was advised by one eminent neurologist to have his head trephined, although he had no scar, no depression of the skull, no typical symptoms of a localized irritation in the brain, no circumscribed pain at any spot, no paralysis, and, in fact, no indications as to just where to trephine or for what. This the neurologist who advised operation acknowledged.

At my request Professor Robert F. Weir, M.D., examined him to determine if he found any indication for trephining, and decided most positively that no indications existed for so dangerous a procedure. The patient disclosed a high degree of esophoria and was treated by me at irregular periods during 1894. Since then he has been too far away from me and too busy in his profession to follow up the eye-treatment as he should have done. He has been greatly improved in his general health and is now busy in his calling. In the frequency of his epileptic seizures he has been markedly benefited (according to the last report). He has not yet been cured of his epilepsy, nor is the maladjustment of his eyes yet perfectly corrected.

CASE XXIV.—Master R., aged 9 years. Referred to me on January 7, 1889, by Dr. J. W. Wright, of New York.

Family History: Both parents are living and healthy. A

brother has severe headaches. Several cases of phthisis and cancer have developed in maternal ancestry. Some phthisis also in paternal ancestry. One paternal and one maternal uncle had epilepsy.

History of the Case: Patient was a delicate child and was circumcised when 18 months old for peevishness, crying, and sleeplessness.

In April, 1881, he had a series of epileptic attacks. The next series followed in October, 1881. Both were very severe and several attacks occurred in succession, chloroform being required to control them. Screaming attacks were also frequent at night and would last for hours. The child was at once put upon bromides and did not have another severe attack for a year, but had numerous attacks of petit mal. A third series of severe convulsions occurred in October, 1884. He was then put upon full doses of bromide and corrosive sublimate and had no severe seizures until March, 1887, but during this time had very frequent attacks of petit mal.

In 1885 severe outbreaks of uncontrollable temper occurred; the child could not be trusted with children, and in temper had been known to throw knives, forks, or anything he could lay hands on. From that time petit-mal attacks have been very frequent, and he has had a series of severe convulsions in March, 1887, December, 1887, August, 1888, and November, 1888. He has taken heavy doses of bromides periodically, but they seemed to exert no special influence.

Eye-defects: Under atropine patient showed: Hypermetropia; O. D., + 0.75; O. S., + 0.50; abduction, 33°; adduction, 6° —; sursumduction, right, 3°; left, 3°; no hyperphoria; esophoria, 6°. Within a week after wearing esophorial prisms he showed, while wearing a 5° prism: Esophoria, 8°; abduction, 2° —.

Treatment and Results: On account of the age and the peculiar temperament of the child, it was found impossible to operate upon him under cocaine. Not even his nurse, who had great influence over him, could persuade him even to have a solution of cocaine dropped in his eye.

As he showed homonymous diplopia with a red glass and at times as much as 20° of esophoria, it was deemed wise by his attending physician and myself to operate under chloroform. He was put under the anæsthetic with difficulty, and both in-

terni were rather freely divided. Great care was taken to avoid
an overcorrection, and, accordingly, the result was not as good
as might have been obtained had I followed my customary
methods and tested the patient's muscles after each tenotomy.

Two days after the operation the tests showed: Esophoria,
0 to 6°; abduction, 6°. Three days later the patient had a
very severe convulsion, preceded by a long aura of terror. This
seizure required the administration of chloroform to control it.
It was followed by eleven light attacks. The child began to
develop considerable latent esophoria within the next month,
and it was becoming evident to my mind that further operative
work on his interni would be demanded. To my great dis-
appointment, the parents then decided to discontinue treatment
for awhile at least, as even the thought of further tenotomies
seemed to greatly disturb the child and his parents. Under
such circumstances very little, if any, benefits could reasonably
be expected from the eye-treatment up to this point, as the case
was one that would undoubtedly require considerable time for
the full solution of the eye-problems encountered. I am informed that subsequently the parents placed the child in a
private institution in England. This case should in no respect
be counted in any inquiry when figuring percentages of results,
as the eye-treatment was abandoned immediately after the first
attempt to rectify his muscles.

CASE XXV.—Miss C., aged 15 years. Referred to me on
April 7, 1892, by Dr. F. H. Olin, of Southbridge, Mass.

Family History: Mother is delicate with pulmonary
trouble. Maternal grandfather died of phthisis. Father has
had severe headaches all his life. His mother and all of his
brothers also have headaches. No chorea, epilepsy, or insanity
in ancestry so far as known.

History of the Case: Patient had spasms in infancy, which
were attributed to indigestion. About four years ago she
began to have light attacks of epilepsy in school. In these she
had an "aura" and momentary lapses of consciousness, but
never fell.

In August, 1891, she had an attack in a store, walked
into the street unconsciously, fell in the street, and was carried
back into the store. She was at once put upon bromides, as
she showed marked indications of another attack the next day.

In September, 1891, she had her second severe fit in a drug-store, and was carried home semiconscious by her family physician. Bromides were pushed for three months and then were stopped for a month.

Mild attacks appeared in January, 1892, and bromides were again commenced.

In March, 1892, she had her third severe attack, followed by another inside of an hour. Both of these were terribly severe.

When I first saw her it was a rare thing for her to go a day without a light attack, and she often had several in a day. All bromides had been stopped a week before. In other respects the patient is well and strong. She is a remarkably bright girl and an accomplished musician.

Eye-defects: Under atropine the patient showed: Hypermetropia ($+ 0.50$); adduction, $28°$; abduction, $4° +$; sursumduction, right, $1° +$; left, $1° +$; right hyperphoria, $0.25°$; esophoria, $1°$. After exercise of the muscles and use of prisms for a week she showed, while wearing $4°$ prisms: Adduction, $45°$; abduction, $1°$; no hyperphoria; esophoria, $5°$. Later she developed considerable latent esophoria and right hyperphoria.

Treatment and Results: The treatment of this case consisted in the wearing of prisms and a graduated tenotomy upon the left inferior rectus and both internal recti muscles. The esophorial operations were performed within the first month, about two weeks apart. During the next month her muscles were tested almost daily, and esophorial prisms were given for suspected latent trouble; but, her tests being rather uncertain, the prisms were removed and the patient sent home for the summer. At that time she showed: Adduction, $50°$; abduction, $6° +$; no hyperphoria; no esophoria. During these ten weeks she went several times from one to two weeks without any attacks, and would then have several light ones, usually during menstruation. She had no severe attack during this period of ten weeks.

July 28, 1892. Her mother reports that during the past five weeks she has had nine slight attacks, four of them during menstruation. In none of these was she unconscious.

September 29th. Patient returned to-day for further treatment, and reports eight light and no severe attacks for the past two months. Her eye-tests were taken daily for a week, but,

as no esophoria disclosed itself, she was sent home for six weeks.
Her tests were: Adduction, 45°; abduction, 5° +; no hyperphoria; no esophoria.

January 10, 1893.   Since last note (twelve weeks ago) the
patient has had five severe and fifteen light attacks.

February 16th.   Patient returned to-day for treatment.
She began to show some right hyperphoria, and prisms were
given to correct it, and increased in strength as fast as she
showed any surplus over her glass.   Within ten days she
showed: Adduction, 43°; abduction, 4°; sursumduction, right,
4° +; left, 0; right hyperphoria, 3°; esophoria, 0.   The left
inferior rectus was then divided, and six days after the operation she showed: Adduction, 45°; abduction, 5°; sursumduction, right, 1° +; left, 1° +; right hyperphoria, 0.25°; no
esophoria.   Prisms were given her for exercising the externi
daily, and she was sent home.

April 19th.   Patient's mother reports that she has had five
severe and eleven light attacks in the past nine weeks, and
that her abduction is 7°.   In November, 1893, her physician
visited my office and reported that this patient " was, in his
opinion, decidedly better than when under the influence of
bromides."   She had few severe fits, longer intervals between
them, and a very marked decrease in the petit-mal attacks had
been observed.   This patient was withdrawn from my care
before the eye-treatment was completed, although I saw her at
irregular intervals for a year and did some satisfactory operative
work on the muscles.   I have reported this case as unimproved,
although I am not sure that decided amelioration or even cure
is not possible in this case.   I have lately written to Dr. Olin
for a report, but I have received no reply, as yet, regarding the
condition of the patient.

CASE XXVI.—Miss D., 30 years of age.

The final case which I shall report here, in connection
with a few remarks of my own, has been selected from my
records of a large number of somewhat similar cases because it
illustrates one point to the oculist more forcibly than any other
case which I have yet encountered,—viz., *that the most competent and intelligent observer never knows how near he may be
to a happy conclusion of his work upon heterophoria, so long
as any remaining trace of abnormal muscular conditions exists
in the orbits.*

In this particular case I had to deal with (1) a most intractable case of epilepsy, (2) one that had been drugged for years, (3) one that had been subjected to the treatment of injection of animal extracts [*a la* Pasteur]. (4) one that had no apparent source of reflex irritation outside of her visual apparatus, and (5) one that for a long time failed to improve under my treatment, in spite of very careful refractive work and apparently satisfactory results from several graduated tenotomies.

If there exist to-day any further necessity of arguing with a few members of our profession who deny that epilepsy is ever produced by eye-strain, this case ought to modify their views.

Family History : The father is in good health ; mother is very neurasthenic, has severe headaches, weak lungs, astigmatism, and hypermetropia. One brother has astigmatism. No hereditary nervous disease as far as known.

This patient, when first seen by me, was a remarkably developed woman. As a child she had always been well and strong. She had perfected herself, at an early age, in swimming and various other forms of athletic exercises. In any feat of strength or agility she always excelled. Later she became quite a remarkable musician. Her menstrual function appeared at the normal age. It has always been unaccompanied by any pain or physical disturbance.

Some years prior to my seeing this patient she developed epilepsy. The seizures, which were of a severe type (grand mal), have persisted at various intervals in spite of all the recognized treatment by drugs. She had been placed under the care of the most eminent specialists here and abroad prior to my seeing her, and later, when drugs had failed to accomplish anything, she had been treated for months by hypodermatics of animal extracts (*a la* Pasteur) by Dr. Gibier, of New York, in addition to the internal administration of various combinations of bromides. During this period of anxiety on the part of her parents and the patient herself she had, from time to time, sought for some possible reflex cause for her epileptic seizures. Her teeth had been found perfect and were remarkably beautiful. No rectal conditions existed that could cause irritation. Her pelvic organs were normal, with the exception of a curvature of the coccyx. A suggestion had been made to her to have the coccyx removed ; but its relationship to the epilepsy seemed too doubtful to justify the operation in the minds of others who were consulted respecting it. She had no kidney disease. She was in no way hysterical, but rather unusually placid and free from nervous excitability.

In the early part of 1894 this patient was referred to me by Prof. A. A. Smith, of New York. As she had never had any symptoms referable to the eyes in her life, both the patient and her parents were naturally skeptical as to the existence of any eye-strain. They consulted me, however, in utter despair, —rather with the hope of finding something wrong in the eyes or eye-muscles than the expectation of doing so. The remaining portion of the history will show what steps were taken toward the happy solution of this complex eye-problem.

At the first examination, March 15, 1894. the patient had $\frac{2.0}{2.0}$ vision in each eye. She tolerated $+ 1.50^s$. She also disclosed: Esophoria, $3°$; right hyperphoria, $\frac{1}{4}°$; adduction, $28°$; abduction, $6°$; right sursumduction, $2°$; left sursumduction, $2°$.

On the following day homatropine was used. The vision in each eye was then brought up to $\frac{2.0}{2.0}$ with $+ 1.50^s$.

Prisms of $1°$ in each eye were given for esophoria. The strength of the prisms was gradually increased (as fast as a surplus was shown) until March 19th. when she showed (while wearing $3°$ prisms, base out): Esophoria, $8°$; no hyperphoria; adduction, $40°$; abduction, $1°$. *The right internal rectus was then button-holed*, with an immediate result of: Esophoria, $1°$; abduction, $8°$.

During the next week the patient showed more esophoria, and prisms were again put on. They were gradually increased as before, until on March 26th her eye-tests were (while wearing $3°$ prisms, base out, over each eye): Esophoria, $10\frac{1}{2}°$; adduction, $42°$; abduction, 0. During this week she had two attacks of grand mal. *All bromides and cerebrin injections were stopped at my orders. The left internus was freely button-holed.* Immediate result: Esophoria. $\frac{1}{2}°$; abduction, $13°$.

From March 26th to April 2d the muscles were exercised with prisms daily; the patient had one grand-mal attack. On April 2d the eye-tests showed: Esophoria. $1°$; no hyperphoria; adduction, $40°$; abduction, $9°$; sursumduction, right, $2°$; left, $2°$.

Between April 2d and 17th esophorial prisms were again given and slowly increased as before. There were no epileptic seizures.

On April 17th the eye-tests (wearing 3° prisms, base out, over each eye) were: Esophoria, 9°; no hyperphoria; adduction, 46°; abduction, 4° +. *Right internal rectus again button-holed.* Immediate result: Esophoria, ½°; abduction, 10°.

On April 19th the patient had one grand-mal attack.

On April 23d I ordered + 1.00ˢ for constant use, as it was deemed wise to take into consideration the latent error of refraction.

Between April 24th and June 15th the patient was tested every other day. Her tests averaged: No esophoria; occasionally left hyperphoria, ¼°; adduction, 40°; abduction, 10°. She had seven seizures, but reported that they were not so severe, and not followed by as much headache and prostration as prior to the eye-treatment.

Between June 15th and 25th a ¼° left hyperphorial prism was given and gradually increased, as patient showed a slight surplus, up to 2°. Two seizures occurred during this time.

On June 26th (wearing a 2° left hyperphorial prism) the eye-tests were: Left hyperphoria, 2°; no esophoria; adduction, 40°; abduction, 8°; right sursumduction, 0; left sursumduction, 4°. Ordered 2° prism, base down, combined with + 1.00ˢ in left eye, as it was deemed wise for the patient to go away for the summer.

As considerable operative work upon the eye-muscles had been done rapidly, it was thought best for the patient to discontinue all eye-treatment for about three months. During this time she could settle down to her new eye-adjustments and become accustomed to the new conditions. She left for the country with instructions to return in September following; to write, read, or sew but little; and to live out-of-doors.

Let us see, up to this time, what has been accomplished (from the eye stand-point) during the first three months of eye-treatment.

1. *A very large amount of latent esophoria has been disclosed and relieved by three graduated tenotomies.* In every instance, except in the last (prior to operation), while the patient was wearing prisms, the abduction fell enough below the normal point (8°) to justify the amount of esophoria disclosed by the patient and to warrant an operative step for its relief. The abduction stands at 8°, and the interni (after three graduated

tenotomies) are able to fuse images over 40° prisms. In other words, both the *interni and externi show normal power.*

2. Left hyperphoria (at times suspected by me in the early tests of this patient) has now begun to disclose itself. The sur-sumduction-tests show: Right, 0 ; left, 4° ; the difference being *just twice the prism that she is wearing* (2°). This warrants us in continuing it; and the patient shows a distinct surplus over the 2° prism (2¼°). In my experience, each degree of hyper-phoria has far greater clinical significance than of exophoria or esophoria. Even fractions of a degree are therefore important.

3. While there has been no marked improvement in the epileptic seizures, the patient is having no more attacks than before the eye-treatment, in spite of the fact that the bromides and the animal-extract injections have been suddenly withdrawn. Otherwise, the patient is as well without drugs as she ever has been for years with them.

4. The full refractive error (exhibited under homatropine) has not yet been corrected. The patient will not yet tolerate more than $+ 1.00^s$ for constant use.[1]

5. Up to this date I had feared that *the use of animal extracts* prior to my seeing this patient had seriously hampered me in my work. Although I had seen, until then, but one case of epilepsy that had been treated by them, I had been seriously alarmed by a sudden and marked diminution of sight that fol-lowed an injection of some animal extract which the patient had been taking without my knowledge. The vision returned in a few days, but the case never did well afterward under my care. So strong was my conviction as to the effect of these animal in-jections upon the treatment of heterophoria that I felt compelled, before accepting this patient, to express great doubt as to good results following my work. Apparently, in this case, however, the results of eye-treatment have not been modified in any way by the use of animal extracts.

6. The most careful questioning of the patient regarding the apparent causes of the various epileptic seizures has thus far failed to establish any relationship between the attacks and imprudence in eating or the menstrual function. They seem to occur without any exciting cause so far as the patient can discern.

---

[1] Later the use of atropine showed some astigmatism, in addition to the hyperme-tropia, that entailed a change of glasses for constant wear.

With these comments as a digression, we are now prepared to continue the study of the clinical history of this case.

The patient returned to the city on September 11th, and reported that she had had four epileptic seizures since leaving the city on June 28th.

Between September 11th and October 22d the patient's eyes were tested frequently, but no more hyperphoria was disclosed. She had, during this time, five epileptic seizures.

On October 23d a graduated tenotomy was performed on the left superior rectus and the prism removed from her glasses. The next day the eye-tests were: Left hyperphoria, $\frac{1}{4}°$; adduction, 40°; abduction, 8°; right sursumduction, 3°; left sursumduction, 3°.

From October 24th to December 11th the patient had six epileptic seizures.

On December 12th some doubt arose in regard to the refraction of the patient, as she tolerated cylinders. To thoroughly eliminate *this* feature of the case a 1-per-cent. solution of atropia was instilled into each eye. Under the atropine the vision was brought up to $\frac{20}{15}$ in each eye with the following correction: O. D., $+ 1.00^{s} \subset + 0.50^{c}$ (axis, 90°); O. S., $+ 0.75^{s} \subset + 1.00^{c}$ (axis, 90°). Glasses were given for constant wear to fully correct this refractive error.

From December 12, 1894, to April 5, 1895, the patient was under almost daily observation, and her eye-tests were perfect,—*i.e.*, no esophoria; no hyperphoria; adduction, 38°; abduction, 8°; right sursumduction, 2°; left sursumduction, 2°.

In spite of her improved eye-conditions and general good health she continued to have about two attacks of grand mal each month. Determined to leave no source of possible reflex irritation uncorrected, I again sent the patient to one of the leading surgeons in New York to have her spine and pelvis examined. He strongly advised against any operation for the slight curvature of the coccyx, as he did not consider it sufficient to cause any reflex trouble.

On April 5, 1895, the patient showed, for the first time in some months, a very slight amount of left hyperphoria, and a $\frac{1}{2}°$ prism was given. This was increased, during the next few days, to a 1° prism. As no more hyperphoria appeared during the next few days, a 1° prism, base up, was combined with the spherocylinder in her right eye.

Her eye-tests on February 6, 1896, ten months from the last record, were: No esophoria; left hyperphoria, 1°; adduction, 38°; abduction, 8°; right sursumduction, 1°; left sursumduction, 3°.

On April 1, 1896, the patient was seen in my office, and her eye-tests were identical with those of April 5, 1895.

The patient has not had a single seizure since the prism was given, over one year ago. She seems in perfect health.

It may be well to draw the attention of the reader to some facts of interest to the general practitioner as well as the oculist outside of the remarkable termination of the eye-treatment of this patient for epileptic seizures.

I would call particular attention to these facts in the clinical history presented:—

1. Both the esophoria and hyperphoria of this patient were almost totally latent. In other words, the majority of oculists would probably have failed to discover it on a casual examination. The patient had learned (by years of practice) to unconsciously adjust for these congenital defects and to simulate a close approach to a manifest orthophoria.

2. Persistent watching of the effects of prismatic glasses judiciously given from time to time, and three graduated tenotomies (performed when the muscular tests warranted their employment), brought about the establishment of, apparently, perfect orthophoria for some months. The patient showed full power of abduction, adduction, and sursumduction.

3. When the eye-tests were persistently normal for some months and every point in refraction had been carefully determined and corrected, I was, naturally, at a loss to explain why the epileptic seizures persisted, and I was led to doubt if some other source of reflex disturbance had not escaped all the physicians who had examined and treated this girl.

4. In spite of this doubt (and in connection with investigations of the spine and pelvis made by others at my suggestion), I still kept up an almost daily watch for some type of latent heterophoria.

In this search I was at last happily rewarded. The one degree of latent hyperphoria (that was at last disclosed) proved to be the remaining source of reflex irritation in this case. Since the day of its discovery no epileptic attacks have occurred. For the first time in many years month after month passed without an epileptic seizure.

This patient presented a very complex eye-problem that was extremely difficult to solve. Very complete details of the treatment of this case have already been published.

Until within a few weeks past she has been totally free from attacks for a period of sixteen months.

During the summer, from causes that are somewhat obscure, she got her digestive apparatus thoroughly upset and had four epileptic seizures.

Her eyes have been examined lately on her return to this city and no material change in her eye-tests has been found. Her glasses are apparently properly focused and adjusted; so that I am at a loss to account for this return of her epilepsy, which I trust will prove temporary.

As she had been under my observation only two years and six months (more than half of which time she had been free from epileptic seizures and observed by me at very rare intervals), I do not consider that the eye-problems are yet thoroughly solved. She is to-day wearing, and has worn for over a year, a 1° prism for left hyperphoria; and some latent hyperphoria still remains to be corrected in this case.

I have reported her case in the table as "decidedly ameliorated," but not as cured.

SUMMARY.

In closing, I think I have proved to the satisfaction of most readers that, out of the 26 cases whose records I have published up to this date, 4 have abandoned treatment almost from its beginning and should not be counted. Of the 22 remaining cases, 10, or 45 per cent., may be considered as well (7 being completely cured and 3 being practically cured); amelioration of the attacks has been afforded by eye-treatment in 9 cases, or nearly 42 per cent.; and no improvement has been observed in 3 cases,—about 13 per cent.

Seven cases completely cured: Nos. 1, 4, 7, 10, 12, 13, and 20 of table.

Three cases practically cured: Nos. 5, 6, and 15 of table.

Nine cases of amelioration: Nos. 2, 9, 14, 18, 19, 21, 22, 23, and 26, of table.

Three cases not improved: Nos. 16, 17, and 25 of table.

Four cases not counted: Nos. 3, 8, 11, and 24 of table.

Total, 26 cases.

Some of the cases reported as ameliorated are still under my observation and may eventually be cured.[1]

CONCLUSIONS.

1. In epilepsy I regard an examination of the eyes (for errors of refraction) and of the eye-muscles (for heterophoria) as the first and perhaps the most important step toward a search for sources of reflex nervous disturbance.

2. No final conclusion should be reached regarding the presence or absence of heterophoria until sufficient time, patience, and skill has been bestowed upon the investigation by one who is thoroughly familiar with and practices the *later methods* for the *determination of latent heterophoria*.

3. All preparations of bromides and other drugs that tend to control the seizures should be withheld, as a rule, from an epileptic patient until all possible sources of reflex irritation have been scientifically sought for and, as far as possible, relieved.

There may be justifiable reasons, to my mind, for *exceptional departures* from this rule; but I wish to emphatically raise my voice here in protest against the prevalent system of drugging epileptic patients from the date of the appearance of the first fit without any attempts being made to ascertain what the causes of the epilepsy may be.

4. I strongly advocate the *employment of atropine in every case* before a final decision is reached regarding the refraction of an eye.

I also believe that in epileptic cases it is wise to insure a full correction of any existing astigmatism (by glasses to be worn constantly in well-fitted spectacle-frames), and as near a full correction by spherical glasses of any latent hypermetropia that is detected as the patient can be made to tolerate, even if

---

[1] The reader is referred to pages 142 and 143 for some important hints relating to eye-work upon epileptic subjects.

atropine has to be instilled at intervals into the eyes of the patient for several weeks to prevent a return of ciliary spasm.

I have observed many cases of chronic epilepsy that have been relieved of all convulsive seizures so long as the full effect of atropine upon the ciliary muscle was maintained.

5. *No promises that absolute cure can be effected by eye-treatment should ever be made to an epileptic;* but it is usually safe for the physician and patient to hope that a radical correction of marked heterophoria and abnormal refraction will eventually be followed by decided and permanent benefits.

6. The results in all cases thus far treated by me seem to warrant the conclusion that at least *90 per cent. of chronic epileptics have been better without bromides,* after a satisfactory correction of their eye-defects, than they ever were when subjected to the influence of drugs. *Some have apparently been cured.* By eye-treatment, up to February, 1897, at least 20 per cent. of my cases have apparently been practically cured of epileptic seizures; but I doubt if so large a percentage of recoveries can be safely predicted *without* medication in any large number of consecutive cases under the best of management by an expert in this line of work.

It should be remembered that a victim to chronic epilepsy who is rendered by any treatment as free from attacks without the bromides as when under their deleterious influence has been *very markedly benefited;* again, that, if a marked diminution of the attacks has been effected, the patient has double cause for gratitude; finally, that, if the attacks are arrested *in toto* without drugs, it is to-day one of the most remarkable facts recorded in medical literature.

7. In cases where negative results have been observed in spite of a satisfactory investigation and correction of marked heterophoria and abnormalities of refraction, I would deem it wise (before resorting to drugs for epileptic seizures) to search for other sources of reflex peripheral irritation (such, for example, as bad teeth, phimosis, rectal or uterine disease, scars, floating kidney, etc.).

Furthermore, the detection of chronic kidney disease, syphilis, organic brain-lesions, and depression of the skull is most important prior to the beginning of eye-treatment or a search for other forms of peripheral reflex disturbances.

8. The treatment of heterophoria by *prismatic glasses*

alone is not curative, nor, in my opinion, are very marked beneficial results to be expected from them. Prismatic glasses are valuable aids, however, in determining the existence and amount of latent heterophoria prior to the radical correction of such defects by graduated tenotomies.

9. *The duration of the eye-treatment of epilepsy varies with the eye-problems encountered.* The establishment of orthophoria is not commonly effected in epileptics inside of one year; and very long intervals of rest between the surgical steps may be demanded and thus extend this limit of time to two or even three years.

For the past five years I have refused to begin work upon the eyes of any epileptic unless I was assured that I could control the patient for a period of at least twelve months, and see the patient at such intervals as I deemed necessary during that period. Experience has taught me in this work that it is much safer to do it slowly than rapidly.

10. Too much stress cannot be laid upon the importance of accurately centering all spherical and cylindrical glasses to the pupils when prescribed. *An imperfectly-fitted frame may prove a source of great injury to a patient.* Moreover, a decentered glass may often lead a careful observer into serious error respecting heterophoria.

11. The objections that are raised before patients by many physicians and some oculists to the performance of a graduated tenotomy upon an eye-muscle "because of its risk, danger of overcorrection, etc.," too frequently cause unnecessary alarm and often prevent patients from obtaining the relief that is within their grasp, and which prismatic glasses have not and will not afford.

The operation, as now performed, does not necessitate any confinement to the house, and all surgical dressings to the eye are avoided.

That the successful operator must possess skill and a high grade of technic (acquired by constant practice in that special line of surgery) is apparent; but I believe that the danger of overcorrection is very small in operating properly for heterophoria, and, in the hands of an expert, an overcorrection is very readily controlled and rectified. I do not recall a case within the past ten years where I have had a moment's anxiety resulting from an overcorrection, although I have in that time performed at least one thousand graduated tenotomies upon the muscles of the orbit.

I think it wise for me to make here a general *résumé* of such points relating to reflex epilepsy as I wish to stand on record.

GENERAL RÉSUMÉ.

(*A*) I think I have shown that the removal of reflex irritation will alter the course of genuine epilepsy. Eighty-seven per cent. of this set of published cases of reflex epilepsy have either been cured completely or markedly benefited by the relief of eye-strain. Other men of high repute have had and published similar results.

(*B*) I have brought forward here some very strong written testimony from physicians of repute and the patients themselves to prove that the histories published by me here are those of genuine epileptics, that the results were as I stated, and that benefit followed the relief of eye-strain after a failure of medicines or diet to control the seizures.

(*C*) I think I have shown that I am not afraid to rest my claims on clinical facts, suppressing nothing and endeavoring to throw all possible light upon the points at issue.

(*D*) I think I have shown that the benefits which these patients have experienced are not due either to stopping the bromides or to simple counter-irritation. If this were so, why do not some of my critics do the same and get the same results? They might get up counter-irritation on some of the large number of epileptics to whom they offer now but little encouragement.

(*E*) I deny, as absurd and untenable, the remarkable claim which has been frequently made,—viz., that the mental effect on a patient from a tenotomy is all that is produced. This is too weak a claim even for argument. Any adversary must be in the last ditch when this is the only loop-hole of escape from clinical facts.

The effects of graduated tenotomies upon the relative power of the eye-muscles, as well as upon the adjustment of the eyes, are too definite and positive (when done with skill) to justify any one in attempting to make the " mental effect " appear more prominent than the actual effect. The latter can be scientifically measured ; the former is mere speculation.

(*F*) I deny the implied claim, made by some critics, that a sort of permanent mental hypnotism is a possible factor in my results. I should be proud to possess any psychological power that could confer health and happiness upon sufferers that I meet; but I must modestly disclaim any such happy endowment, and give to nature alone the credit of re-establishing herself after her burdens have been removed.

(*G*) I think the total percentage of epileptics who suffer from eye-strain as an important factor is very large, after first deducting from the total number the comparatively small number of cases that owe their epileptic seizures directly to some organic lesion of the brain or to a depression of the skull.

Almost all chronic epileptics give a history of falls that have sometime injured the head in some way. Few of these cases, however, have enough depression of the skull to make trephining imperative, and in every such case the injury must have preceded any epileptic seizures to make it probable that the fits were the direct result of the injury.

(*H*) The enormous percentage of complete and practical recoveries (in this set of twenty-two cases reported here, where eye-strain was relieved) is much larger, I think, than any one can reasonably hope to obtain in any larger number of cases, even when the oculist is particularly skillful in solving the complicated eye-problems of epileptics, and has had a wide experience in this field. Such a percentage as reported here is vastly greater than I have ever claimed or even hoped for in epileptics.[1]

(*I*) I think I have shown that eye-strain can exist without any eye-symptoms, and that " pain in the occiput and nape of the neck " need not necessarily exist, although common in such cases.

(*J*) I think I have explained quite fully why epileptic patients may have temporary relapses after good results from eye-treatment, without in any way justifying invidious criticism upon the treatment or the permanent benefits that might have been uninterrupted if the patient had avoided new sources of reflex irritation.

(*K*) I would impress the reader with the fact that any

---

[1] Epilepsy is regarded by most authors of repute as almost an incurable disease. If any one could *positively cure all cases*, the entire hotel accommodations of New York City would not be sufficient to hold the epileptics that would apply for relief.

marked amelioration of epileptic seizures (in violence or frequency) without drugs is a great step in advance of previous methods of medication (even if the cure is not complete).

(*L*) I would again impress upon the medical profession the extreme difficulties of eye-treatment of chronic epileptics, and the necessity of long-continued and patient watching for latent errors of adjustment, before operative work is discussed or attempted. Moreover, it is important that the oculist be familiar with the new methods, and that the patient be sufficiently intelligent to realize the importance of details and to be persistent in the treatment until perfect adjustment of the eyes is established. I received the following letter from a country practitioner that tells its own story :—

DEAR DR. RANNEY : I have read your articles on the eye-treatment of epilepsy with great interest. I know little about eyes, but I have an epileptic patient that cannot be cured. I am tempted to cut an eye-muscle and see if it does any good. Respectfully yours,

DR. ——.

(*M*) I think it can be shown that many eminent medical men, who bitterly opposed, in years past, the views advocated here, have been forced, at last, to give some recognition to the eye-treatment of nervous diseases. They are not yet enthusiasts, perhaps, nor are they all skillful in the work, but they cannot afford to longer oppose, with manifest bigotry or intolerance, the clinical facts that have been brought to their notice. Quotations can easily be made from many of the latest text-books on the eye and on nervous diseases to show how much attention has been given to eye-muscles (in contrast with text-books of the past).

(*N*) I would again raise my voice in protest against treating any form of nervous disease with drugs (especially the more intractable types, such as epilepsy, insanity, chorea, and neuralgias) until a very careful and intelligent search has been made for all reflex causes. It may take time to do this, and it may involve some expense, but it is often the shortest and surest way to effect a cure. It is scientific rather than an empirical and purely speculative method of stopping symptoms instead of the cause.

I now present a tabulated summary of the twenty-six cases of epilepsy reported by me. They were treated entirely through

the correction of errors of refraction and anomalies of adjustment of the eyes.   I have reproduced the original records[1] of
January, 1894, in order that the reader may contrast the four
right-hand columns with each other and thus see the progress
of each patient during the last three years.   The other columns
give the reader information that is easy of reference respecting
the history of the case, the refractive errors that existed, and
the treatment of the eye-muscles.

This somewhat lengthy chapter on epilepsy finishes my
contribution to an attempt to see justice done to a new and
rapidly growing method.   I do not hope in this world to see the
lion and the lamb lie down in peace together; but I have not
yet lost faith that, in time, careful, honest, conscientious, and
painstaking work will bring the reward of appreciation and respect even of those who differ with me.

If it be true, as I think time will surely demonstrate it to
be, that epilepsy is commonly a reflex rather than an evidence
of organic disease, the medical profession will have to be educated to discard therapeutics until the *radical elimination of all
existing reflexes* has been attempted by skillful men and the
epileptic attacks still persist.   Such an education, of necessity,
takes time.   Moreover, those who are truly skilled in special
lines of medical work that bear upon the treatment of epilepsy
are fewer in number to-day than they surely will be in the
future.   Even in the medical staffs of public institutions it is
very difficult to-day to find any one who is willing to devote the
necessary time and trouble to epileptic patients.   It is so much
easier to give bromides, and the prognosis (according to text-
books) seems so bad to most medical men, that these poor
victims, helpless and hopeless, are either given up to routine
drugging or are at last turned out without drugs into epileptic
colonies, where they can have their fits with less personal
humiliation and without social ostracism.

The late investigations that have been made relative to
abdominal reflexes from floating kidney bid fair to aid in some
cases of intractable epilepsy.   I have lately seen two cases
where extreme heterophoria had existed and been satisfactorily

---

[1] The records of these cases were first published by me in 1894 in the New York
Medical Journal.   Since that date sufficient time has elapsed to show whether there was
any permanency to my results or whether the patients relapsed to their former epileptic
condition.

corrected by me without a complete cessation of the epileptic seizures. I was at a loss to discover any cause for these sporadic attacks, especially as the nose, teeth. pelvic organs, rectum, and ears had been examined and found to be normal. At last I suggested to these patients the advisability of being examined for a suspected floating kidney. The persistence of digestive disturbances, difficulty in sleeping upon one side, and some bladder disturbance seemed to me a cause for suspicion of some abdominal reflex. One of these patients has lately been operated upon for a displacement of about six inches of the right kidney; the other has an equally marked displacement, but has not yet decided to attempt its fixation. The effects of this work upon the epilepsy has not yet been determined.

A Detailed Summary of the Treatment of Twenty-six Cases

Case No.	Date First Examination	Initials	Age and Condition	Previous Treatment and Results	Refractive Errors			Muscular Anomalies	Ocular Treatment Employed	Drugs Administered by Us	Number of Attacks While Taking Bromides
					H.	M.	A.				
1	July 30, 1892.	Mr. B.	26 yrs.; married.	Bromides for 2 months; negative results.	+0.50 +0.50			Eso. Exo. Hyp. ? Additional latent esophoria discovered.	Three graduated tenotomies upon the interni.	None.	From two to ten daily.
2	Sept. 2, 1892.	Mr. F.	27 yrs.; single.	Bromides for 15 years; negative results.	−0.50		+0.50 −0.50	6° .... 1° Additional latent heterophoria discovered.	Three graduated tenotomies for relief of esophoria and hyperphoria; glasses for constant wear.	None.	Four severe seizures in 7 months prior to eye-treatment.
3	May 5, 1893.	R. G.	10 yrs.	Bromides for a time; negative results.	+1.75 +1.75 More hypermetropia found later.			Double convergent strabismus.	Two graduated tenotomies upon the interni; hypermetropic glasses.	None.	Extremely frequent; often during each day.
4	Jan., 1896.	Mr. H.	43 yrs.; married.	Bromides and other drugs for 24 years: negative results.	+2.50 +2.50			4° .... Additional latent esophoria discovered.	Two graduated tenotomies upon the interni; hypermetropic glasses.	None.	About four severe attacks during each year.
5	Mar. 1?, 1890.	Mr. H.	24 yrs.; single.	Bromides in all possible combinations for 24 years; negative results.	+4.00 −1.00		+0.50	11° .... Additional latent esophoria discovered.	Two graduated tenotomies upon the interni; full correction of astigmatism by glasses.	None.	Two attacks during the year prior to eye-treatment.
6	Nov. 27, 1888.	Mr. S.	19 yrs.; single.	Bromides in enormous doses, with chloral, for many years.	+0.50 +0.50			4° .... ½° Latent esophoria and hyperphoria disclosed themselves to a high degree.	Graduated tenotomies for esophoria and hyperphoria.	None.	Thirty-four days during the year prior to eye-treatment had been attended with a series of convulsions.
7	April 6, 1889.	Mr. S.	16 yrs.; single.	Bromides at intervals, but in small doses	+1.00 +1.00			5° .... Latent esophoria was disclosed to a very high degree.	Graduated tenotomies; glasses for reading.	None.	Seizures somewhat irregular, about four a year.
8	Oct. 28, 1892.	Mr. O.	28 yrs.; single.	Bromides for 3 years.	+0.50	−2.50	−0.50 +0.50	.... 16° 2° Crossed diplopia; hyperexophoria.	Three graduated tenotomies; full correction of refraction by glasses.	None.	Three attacks during year while under bromides.
9	May 2?, 1888.	Miss S.	13 yrs.;	Bromides for 3 years; negative results.	+1.50 +1.75		+0.75 +0.75	10° .... A high degree of latent esophoria disclosed itself later.	Three graduated tenotomies; glasses for constant wear.	None.	About two severe seizures each month.

OF CHRONIC EPILEPSY BY CORRECTION OF EYE-STRAIN ALONE.

NUMBER OF ATTACKS ON STOPPING BROMIDES.	RESULTS OF EYE-TREATMENT TO JAN. 1, 1894.	REMARKS ON CASE (JAN. 1, 1894).	RESULTS OF EYE-TREATMENT (NOV., 1896).	REMARKS ON CASE (NOV., 1896).
One hundred and six attacks during the first fourteen days.	No attack for more than 12 months.	Patient has been actively engaged as a skilled workman on machinery during the period covered by eye-treatment.	COMPLETE CURE. No attack of any kind since last reported attack (about 4 years ago).	This case seemed very unpromising to me when eye-treatment was commenced. He wears no glasses.
Not determined.	Only one light attack since first visit (September, 1892).	This patient has used his eyes on an average of six hours per day at book-keeping: he has entirely recovered from chronic dyspepsia of years' standing.	GREAT AMELIORATION. Eight attacks only in 4 years—some very light.	The attacks have been reduced about 75 per cent. This patient would probably have been completely cured of epilepsy if he could have followed an out-of-door occupation. His work as a book-keeper entails too much strain upon his eye-muscles. His dyspepsia has been cured.
Not determined.	As far as known, only one attack has occurred since the first visit.	This patient has made a remarkable recovery from partial idiocy.	Patient abandoned treatment after only about one week of my care. The fits have returned, and idiocy also.	This patient lost his glasses in a river. His family refused to replace them or to follow my directions. In spite of wonderful improvement the treatment was abandoned.
Not determined.	No attack for about 7 years.	This patient has been using his eyes constantly for years without any asthenopia or headache.	COMPLETE CURE. No fits in nearly 10 years.	This case had withstood all medicinal treatment for 24 years. His recovery was rapid and permanent after eye-treatment.
Several times as many as when the bromides were given.	Only one slight seizure during the past 3 years.	The effects of bromides upon the mental condition of this patient were alarming. He has entirely regained his mental and physical health, and has lately married.	PRACTICALLY CURED. Only three attacks in nearly 6 years, due to causes that might easily excite a convulsion in a healthy subject (see late record).	Every attack has been directly produced by an over-indulgence in wine or rich food. The slightest change in the axis of his cylindrical glass is apt to create serious nervous disturbance in this patient.
This experiment was never deemed safe.	Two years and three months without an attack, and only one slight seizure in nearly 3 years.	This patient was afflicted with epileptic mania, and was at one time about to be committed as an incurable epileptic to an asylum. He required a room padded with mattresses while his seizures were active.	PRACTICALLY CURED. Only one series of attacks in past 4½ years, following a fainting attack caused by extreme pain and while living in a high altitude.	This case properly belongs, in my opinion, to the class of complete cures. The only attack that is reported during the long interval of 4½ years was produced by the pain of two enormous boils. He has had no attacks for 14 months.
Not in excess of number under bromides.	No attacks for past 2½ years.	The family history of this case shows that eye-defects were inherited by the patient. Serious nervous conditions had developed in the father and a brother.	COMPLETE CURE. No attacks reported in over 5 years, during which time he has been a student.	A brother of this patient has been an inmate for years of an institution for the education of the feeble-minded. The father has shown symptoms of insanity.
Not determined.	Not determined; patient returned to the bromides contrary to my advice.	This patient became alarmed because he had some seizures after stopping the bromides, and abandoned the eye-treatment. The progress of the eye-treatment had been more than satisfactory to me.	Patient never returned for further eye-treatment, nor did he follow my instructions.	In figuring percentages of results this case should not be counted.
Continuous epilepsy that endangered life, within 24 hours.	Epileptic seizures somewhat less than when under the influence of bromides; the physical condition of the patient is greatly improved.	This is one of the cases where the marked improvement in the patient is not as clearly shown by the numerical decrease in the epileptic seizures as by the changes in the patient herself. She is physically a different being than when taking bromides.	AMELIORATION OF ATTACKS. No late report has been obtained either from the physician or the family. I have not seen the patient for 3 years, and do not now know her address.	The treatment of this case was begun and finished before the instruments of precision of the present day were sufficiently perfected to enable the oculists to do creditable work upon the complex ocular problems of epileptics.

A DETAILED SUMMARY OF THE TREATMENT OF TWENTY-SIX CASES OF

Case No.	Date First Examination.	Initials.	Age and Condition.	Previous Treatment and Results.	Refractive Errors.			Muscular Anomalies.	Ocular Treatment Employed.	Drugs Administered by Me.	Number of Attacks While Taking Bromides.
					H.	M.	A.				
10	Oct. 22, 1890.	Mr. F.	40 yrs.; married.	Had never taken any bromide salts.	+1.50 +1.50	....	....	Eso. Exo. Hyp. 6O Homonymous diplopia prior to tenotomies.	Four graduated tenotomies; spherical glasses for constant wear.	None.	Had never taken bromides.
11	Mar. 27, 1891.	Mr. D.	22 yrs.; single.	Had taken bromides for years, but had abandoned them for 6 months prior to eye-treatment.	+0.75 +0.75	....	....	5O Latent esophoria existed.	Two graduated tenotomies upon the interni; glasses for his refractive error.	None.	Frequent severe seizures.
12	Mar. 1, 1893.	Mrs. G.	30 yrs.	Bromides for 3 or 4 years; no benefit derived. Serious mental and physical effects were apparent.	+1.00 +1.50	....	+0.50 +0.50	2O Latent esophoria disclosed itself to a marked degree.	Two graduated tenotomies upon the interni; correction of the refractive errors by glasses.	None.	Paroxysms of continuous epilepsy at intervals that would last from 24 to 48 hours.
13	April 8, 1893.	Mrs. W.	30 yrs.	Bromides seemed to exert no influence upon the attacks.	Absolute emmetropia (even under atropine).			7O The esophoria was totally latent.	One graduated tenotomy upon the right internal rectus.	None.	Exact record not kept by family Several occurred during the first week that I personally controlled the case.
14	Mar. 2, 1893.	Mr. S.	19 yrs.; single.	Bromides for 7 years; negative results.	Absolute emmetropia (even under atropine).			7O The esophoria was almost totally latent.	One graduated tenotomy of right internal rectus.	None.	An attack about every 14 days.
15	Sept., 1893.	Mr. P.	26 yrs.; single.	Bromides for some years; negative results.	Emmetropia.			7O Additional latent esophoria.	Three graduated tenotomies upon interni.	None.	Six severe convulsions during the year that preceded the eye-treatment.
16	Mar. 23, 1893.	Mr. H.	18 yrs.; single.	Bromides for some years; negative results.	+2.50 +1.00	....	....	Double divergent strabismus; apparently a right hyperphoria also.	Two graduated tenotomies upon the externi; full correction of refractive error by glasses.	None.	A severe convulsion each month; from two to fifteen attacks of petit mal daily.
17	Dec. 28, 1891.	Miss F.	17 yrs.	Had never taken the bromide salts.	Absolute emmetropia (under atropine).			3O Some latent esophoria was disclosed.	One graduated tenotomy upon the right internal rectus.	None.	Had never taken bromides.
18	Oct. 13, 1890.	Miss J.	22 yrs.	Bromides for years; attacks not arrested, but decreased in number.	+1.00 +1.00	....	....	18O 5O+ Crossed and vertical diplopia.	Graduated tenotomies upon both externi and superior rectus; full correction of refraction by glasses.	None.	About four severe attacks each year.

CHRONIC EPILEPSY BY CORRECTION OF EYE-STRAIN ALONE (cont'd).

NUMBER OF ATTACKS ON STOPPING BROMIDES.	RESULTS OF EYE-TREATMENT TO JAN. 1, 1894.	REMARKS ON CASE (JAN. 1, 1894).	RESULTS OF EYE-TREATMENT (NOV., 1896).	REMARKS ON CASE (NOV., 1896).
Only one severe fit; uncontrollable attacks of nausea accompanied by symptoms of petit mal.	No convulsive seizures for over 2 years; only one attack of nausea during year 1893.	The paroxysms of nausea that formerly lasted a week, and were closely allied to attacks of petit mal, have been greatly modified, and are now very infrequent.	COMPLETE CURE. No epileptic seizure for almost 5 years. One attack of nausea in 1895 without loss of consciousness.	The recovery of this patient seems to be complete in spite of the failure of drugs to control either the epilepsy or the attacks of nausea. He has to use his eyes almost constantly in his vocation.
One hundred severe attacks during past 6 months; twenty-five attacks of petit mal often during 24 hours.	Epileptic seizures were markedly reduced in number prior to the sudden death of the patient.	This patient discarded his glasses contrary to instructions. He was found dead with a wound on the forehead, supposed to be due to falling upon a stone when seized with an epileptic attack.	Found dead *with his glasses in his pocket.*	This patient entirely disregarded my instructions as to the use of hypermetropic glasses. This may have caused the convulsion that that led to his untimely end.
Patient did not dare to abandon bromides until after the second tenotomy.	No attack during the past 5 months.	Frequent and uncontrollable hysterical attacks that occurred prior to the eye-treatment are things of the past. The patient gained 28 pounds in weight within 3 months after the withdrawal of the bromides. Her mental powers have been perfectly restored.	COMPLETE CURE. No epileptic seizures during the past 3½ years.	This case appeared to me to be almost a hopeless one when first seen. Her physical and mental condition was alarming. The letter of Dr. Hedges and the report of her husband both confirm this statement.
Not materially altered.	No attack for past 9 months.	This patient was found to be absolutely free from refractive errors. Her esophoria was also totally latent. The solution of this problem was effected by a judicious use of prismatic glasses.	COMPLETE CURE. No attack of any kind for 3½ years.	This case seemed a desperate one at the beginning of my work. She had no error of focus, and but one graduated tenotomy was required to complete the eye-treatment and to restore perfect health.
Withdrawal of bromides for 4 months did not affect frequency of epileptic attacks.	Patient has passed 6 months without an attack since the operation; only two attacks in 13 months.	A terrible accident occurred to this patient. While in a fit he overturned a lamp and was burned so that his life was despaired of. Since then the eye-treatment has been suspended until lately.	DECIDED AMELIORATION. This patient continues to have attacks occasionally, but has never returned to bromides.	Eighteen attacks have occurred in 3½ years, some slight. This is about one-fourth of the number that his average (while under bromides) would have aggregated in the same time.
Not determined.	No attack for a period of 11 months during 1893.	This patient has been entirely relieved of chronic constipation of many years' standing (as a result of his improved nervous tone).	PRACTICAL CURE. Two or three light seizures have followed severe gastric upsets during the past 3 years.	This patient indulges in very violent forms of athletics constantly as a pastime. He is irregular in sleeping and imprudent in eating. He is a book-keeper, and uses his eyes constantly.
Not determined.	Four months without a convulsion; attacks of petit mal much less frequent.	The eye-conditions of this patient are as yet only partially solved. This case is one of the most difficult cases of heterophoria that I have ever seen.	NOT IMPROVED. The fits and mental impairment of this patient were not markedly improved when last report was received.	This patient was too sluggish mentally to give intelligent eye-tests. I regard this case as one of organic brain disease. I have not seen the patient in nearly 2 years.
About four severe convulsions during each year.	Results negative: patient continues to have attacks every 4 or 5 months.	The improvement in the heterophoria was marked in this patient, but no material change in the frequency of the attacks followed.	NOT IMPROVED. This patient abandoned treatment about 2 years ago.	The probability of a reflex pelvic cause in this case seemed great. No examination was made while under my care.
About one severe attack each month.	Only three attacks in 13 months.	This patient has been enabled to dispense with a constant attendant. She goes to places of amusement, halls, etc., and is regarded as an invalid no longer by her parents or friends.	DECIDED AMELIORATION. This patient still has occasional seizures, but she requires no attendant, and is able to work. She has never completed the eye-treatment.	The parents of this girl decided to abandon eye-treatment before I deemed it wise. When I last saw the patient she looked like a different being than when she first came under my care.

A DETAILED SUMMARY OF THE TREATMENT OF TWENTY-SIX CASES OF

Case No.	Date First Examination	Initials.	Age and Condition.	Previous Treatment and Results.	Refractive Errors. H.	M.	A.	Muscular Anomalies.	Ocular Treatment Employed.	Drugs Administered by Me.	Number of Attacks While Taking Bromides.
19	Mar. 10, 1893.	Miss D.	12 yrs.	Bromides for 18 months; physical results unsatisfactory.	+2.75	+3.25		Eso. Exo. Hyp. 20° A high degree of latent esophoria disclosed itself.	Graduated tenotomies upon the interni; glasses for constant wear.	None.	Attacks arrested for 18 months at one time.
20	June 15, 1893.	Miss D.	30 yrs.	Bromides for over 5 years.	+1.50	+1.50		90 A high degree of latent esophoria.	Two graduated tenotomies upon internal recti; +1.00 s. glass for constant wear.	None.	Ten severe seizures in 5 years.
21	Feb. 17, 1892.	Mr. T.	22 yrs.; married.	Bromides in heavy doses for past 10 years.	+0.50 ... +0.50	+0.50 ... +0.50		7° This patient has shown an approach to double convergent squint prior to my tests.	Both interni had been operated upon prior to my examination of the patient. One graduated tenotomy was performed by me.	None.	An average of six severe fits each year, with some petit-mal attacks.
22	Oct. 24, 1893.	Miss R.	22 yrs.	No bromides.	+1.25	+1.25		15° 90 Extreme nystagmus when either eye is covered.	Graduated tenotomy upon left superior rectus.	None.	Had never taken bromides.
23	Oct. 21, 1893.	Mr. M.	31 yrs.; single.	Bromides for past 12 years.	+0.50	+0.50		7½° Unconquerable diplopia.	Three graduated tenotomies upon the interni.	None.	Four severe fits during past year.
24	Jan. 7, 1889.	Master R.	9 yrs.	Bromides in large doses at intervals for some years; negative results.	+0.50	+0.50		6° Additional latent esophoria.	Two graduated tenotomies upon the interni, under chloroform.	None.	A series of convulsive seizures at irregular intervals. Very frequent attacks of petit mal between the convulsive outbreaks.
25	April 7, 1892.	Miss C.	15 yrs.	Bromides for 6 months; results not satisfactory.	+0.50	+0.50		6° 4° Considerable latent esophoria and hyperphoria existed.	Graduated tenotomies upon both internal recti and left inferior rectus.	None.	Very severe seizures at irregular intervals. Petit-mal attacks very frequent, often several during each day.
26	Mar. 15, 1891.	Miss D.	25 yrs.	Bromides; animal extracts; negative results.	+1.50	+1.50		8° 1°	Graduated tenotomies upon both interni.	None.	Very severe attacks of grand mal about once in each week.

Summary: Completely cured: Cases 1, 4, 7, 10, 12, 13, 20. Amelioration: Cases 2, 9, 14, 18, 19, 21, 22, 23, 26.

CHRONIC EPILEPSY BY CORRECTION OF EYE-STRAIN ALONE (concl'd).

NUMBER OF ATTACKS ON STOPPING BROMIDES.	RESULTS OF EYE-TREATMENT TO JAN. 1, 1894.	REMARKS ON CASE (JAN. 1, 1894).	RESULTS OF EYE-TREATMENT (NOV., 1896).	REMARKS ON CASE (NOV., 1896).
Four severe convulsions in 12 weeks prior to eye-treatment.	Five attacks during past 9 months; seizures much less severe than formerly.	This patient has not yet been observed for a sufficient time to speak definitely about results. Her parents and physician regard her as very much improved by eye-treatment.	SOME AMELIORATION. The patient has not had the eye-treatment completed.	The slow approach of the menses has been an important factor. I think, in causing the attacks during the past 2 years. The eye-treatment has been unfortunately postponed for 2 years or more.
Two severe fits one month apart.	No epileptic seizure for over 7 months.	This patient has steadily filled a clerical position that involved a constant use of the eyes during the period of treatment in my office.	COMPLETE CURE. No attack of any kind during past 3½ years.	This young lady has had to use her eyes constantly as a book-keeper, yet she has been perfectly restored to health by eye-treatment.
Not determined.	During the past year the patient has had six seizures.	The eye-treatment seems to have accomplished as much as large doses of bromides did thus far; apparently the epileptic attacks are growing still less frequent. The physical condition of the patient is very much improved.	SOME AMELIORATION. This patient is very apt to have attacks from gastric upsets. He had an injury to his head years ago that may be a factor in his epilepsy.	This patient had had two operations done upon his interni (before I saw him) by an oculist in Maine. I was therefore embarrassed in my work from the onset from a lack of knowledge of the original conditions.
Almost daily attacks of petit mal.	Attacks less than before operation.	The treatment of this case has not progressed far enough to justify any marked improvement in the epileptic seizures. Extreme heterophoria still remains.	DECIDED AMELIORATION. The actual number of seizures is less than before eye-treatment, and nystagmus is totally arrested unless the right eye be covered or closed.	This case was one of terrific nystagmus. She had been operated upon in infancy for double cross-eye: hence the eye-tests have been very unreliable, and the results even better than I at first hoped for.
Has never dared to abandon the bromides.	No attack since the first graduated tenotomy (nearly 4 months).	In spite of the sudden withdrawal of the bromides, the patient reports a very decided improvement in his general physical condition.	DECIDED AMELIORATION. Occasional seizures still occur, but the patient's physical and mental condition is greatly improved.	This patient has been enabled to return to active work as a minister of the gospel.
Not tried.	Negative.	The patient was withdrawn from my care before the results of eye-treatment could be determined. Marked latent heterophoria remained uncorrected.	This patient has been confined in a private retreat for the past 4 or 5 years.	The marked mental derangement of this boy (at times) and his violent temper made it impossible to carry out a system of eye-treatment with any hope of success.
....	This patient has about as many attacks as when under the influence of bromides.	The correction of the existing heterophoria in this case is probably imperfect. Further operative work will doubtless have to be done before orthophoria is established.	NOT IMPROVED. This patient was withdrawn from my care by her parents in spite of the protests of the family physician.	This was a complicated and difficult eye-problem to solve, and the parents were not content to wait for results. I still feel sure that good results would have been manifested later had the eye-treatment been completed.
Same as when under bromides.	....	....	DECIDED AMELIORATION. This patient passed 16 months without attacks of any kind.	I have strong hopes of yet making the results of this case a permanent and complete cure.

Practically cured: Cases 5, 6, 15. Not improved: Cases 16, 17, 25. Not counted: Cases 3, 8, 11, 24.

CHAPTER VIII.

THE EYE-TREATMENT OF NERVOUS PROSTRATION AND INSANITY.

THE conditions of nervous prostration and insanity seem to be so closely related, in many instances, that the author has deemed it wise to include the records of a few interesting cases of these types in one chapter.

Many victims of insanity, especially in some of the milder types, clearly show, by their clinical histories, that the mental disturbance did not develop while in perfect health, but that it was preceded by a decided nervous debility which had for a variable period of time been recognized by the patients themselves, their immediate families, and friends.

Irrespective of the views that are advanced to-day regarding the organic changes which are encountered in the brain itself or its envelopes in the various types of insanity, it must be admitted, I think, by all medical men having a large experience with the insane, that remarkable and complete recoveries sometimes do occur in patients that apparently for a time gave little basis for a hopeful prognosis.

The clinical histories of two cases that I shall incorporate in this chapter are of this type.

Neither of these seemed to me, at first, to be cases that justified much hope, if any, of ultimate recovery; they presented symptoms that pointed, in the strongest way, toward the existence of organic cerebral disease; they were hard to control by experienced attendants; they had not been benefited (prior to consulting me) by medication, although they had been treated for some time by men of reputation; one was in extremely bad physical condition, so that the probable duration of life seemed to be near its close; and, in each instance, the eye-treatment was begun (purely at the solicitation of friends) rather as a last hope than from faith in its curative results.

One of these cases presented many of the typical symptoms of paresis,—a disease that is generally accepted by the physician as a hopeless malady.

Some years ago, when Weir Mitchell first brought the

" rest cure" into prominence as a means of restoration to health
of cases of general physical debility and nervous collapse, the
happy results obtained by this method of treatment (in which
drugs formed no essential factor) led to its general adoption by
medical men as a step in advance of all previous methods.

It cannot be denied that by this method of treatment thousands of sufferers from nervous prostration have been greatly
benefited or restored to health ; but they have had to bear with
patience its many unpleasant features for months at a time ;
have had to voluntarily isolate themselves from their families.
without even written communications with loved ones; have
been confined to bed ; have been fed under explicit directions ;
have had a trained nurse to direct them ; have had massage and
electricity as a substitute for exercise ; and have in *no way been
allowed to use their eyes* for reading, sewing, playing games, or
other amusements.

The particular factor in this so-called " rest treatment "
that I have put in italics in the preceding paragraphs has not
been given, perhaps, sufficient credit for its results upon this
class of patients.

It is essentially a *rest treatment to vision* as well as to
muscles.

It practically reduces to a minimum in these patients all
leakage of nerve-force from eye-strain. It allows of a fresh
accumulation of nervous energy, because it arrests (as far as it is
possible to do without direct aid to the eyes) excessive expenditures of nerve-force, in those patients who are unconscious
victims to errors of refraction or anomalies of the muscles of
the orbits (heterophoria).

That the distinguished neurologist, after whom this rest
treatment is commonly named and by whom it is extensively
employed, is not indifferent to this view is shown. I think, by
the fact that he to-day has the eyes of many of his patients very
carefully examined during or prior to his rest treatment.[1]

Possibly he has found that his cures are more rapid and
permanent, when eye-strain is corrected early, than in years past,
when the testing of the eyes of neurasthenic patients was less
frequently, and certainly less thoroughly. done, even by oculists
of repute.

---

[1] I base this assertion upon the statements made to me by patients who have been
under his care.

One of the cases reported in this chapter affords a happy illustration of my view :—

After three months or more of the strictest rest treatment she was discharged by the aforesaid neurologist as cured. The improvement lasted only a few months. Gradually all the old symptoms of nervous prostration returned. Her physical and nervous condition became, within one year, as bad as, if not worse than, before the rest treatment.

A correction of a latent heterophoria without any rest treatment restored her to health, and the cure has remained permanent for many years.

To the thinking mind this case suggests the following pertinent inquiry :—

Could not this patient have been spared the confinement; the isolation from husband and children ; the large expense entailed by absence from home, trained nurse, etc. ; and one more year of suffering had the eyes been examined earlier ?

Another case reported in this chapter illustrates, perhaps with equal force, the failure of a somewhat similar experience. Here heterophoria proved to be the exciting cause of peculiar and intractable nervous manifestations in a male, and its correction effected a complete cure.

It is not the author's intention, in this chapter, to discuss in detail either insanity or neurasthenia in their varied types and manifestations. The limits of this small volume would preclude it, if otherwise it seemed to the author judicious to attempt it. It is sufficient to bring forward here (purely as an object-lesson) a few clinical histories that are of great interest because they shed light upon a much neglected factor in mental as well as physical collapse.

Such cases as these are not without parallel in the experiences of other medical men who have devoted themselves to the study of heterophoria. I find in my case-books several letters from physicians that narrate equally startling results from the relief of eye-strain in cases of extreme mental disturbances. Furthermore, similar results have been published by others, from time to time, within the past decade, without apparently attracting the amount of attention that such results should warrant.

I feel sure that in the future more attention will be paid to the eyes of inmates of our insane-asylums. In the past, as far

as I can learn by conversation with medical directors of such
institutions, little, if anything, has been attempted in this line.
Even the *refraction of insane patients* has not been generally
tested under atropine in such institutions; nor are any *systematic records kept of eye-examinations of inmates*, as far as I
know, in any insane-asylum of America or Europe.

I have at times pleaded with several medical superintendents
of public institutions of this kind for an attempt to obtain and
publish a *complete set of eye-tests* (by the new methods) of all
the insane patients under their care; but I have been met with
the statement that the resident medical staff was not competent
to do the work, and that the labor of doing it was too great to
undertake without a special appropriation to meet the expense
of employing competent aid in making the investigation.

A case is reported by my friend, Dr. G. T. Stevens,[1] where
acute mania was arrested almost at once by the employment of a
few drops of eserine in the eye, in order to aid the accommodation that was temporarily paralyzed, and to stimulate pupillary
contraction.

In one of the cases reported by me the use of atropine in
the eye to relax ciliary spasm caused an immediate restoration
of the patient to health. His eyes had never been regarded as
a factor in his nervous disorder.

CASE I. *Symptoms Closely Resembling Those of Paresis.*—
Mr. B., age 37; clerk; married. Patient was referred to me
by Dr. T. J. Martin, of Buffalo, on February 20, 1895.

Family History: Father died of apoplexy at 79; mother
died of Bright's disease at 69. Several paternal relatives died
of paralysis. No hereditary tendencies to phthisis, headaches,
or nervous diseases. Has one son, 11 years old, who is strong
and well.

History of the Case: For the past two years this patient
has shown great irritability in his business. In May, 1894, he
began to have pains in the occiput and trouble in his speech.
The pains subsided in June, but his speech continued to be
troublesome.

In October, 1894, he had a sudden collapse of business
ability,—sending goods to wrong places and misapplying checks

[1] Functional Nervous Diseases, D. Appleton & Co.

18

and money. A loss of memory for events and names developed. Gradually the condition of somnolence appeared; so that he would drop to sleep suddenly.

For the past three months he has taken a course of galvanism. He can now tell names and remember events from day to day, but he has lately been doing strangely-erratic things in the line of purchases and expenditures. The day before leaving Buffalo he went to all the banks trying to cash checks and bought a number of horses, a revolver, etc. He is constantly hunting through the drawers and closets in his house. He will steal and hide sweets and cakes from his own table; he is utterly irresponsible, very hard to control, and imagines that he is very wealthy.

After reaching New York in charge of a male nurse he became more violent than usual, and very difficult to control. He accosted women on the street until a policeman was on the point of arresting him; tried to enter different rooms in the boarding-house; put on his clothes incorrectly,—his drawers over his pants, etc.; washed his feet in the slop-jar; and resisted all attempts to control him, etc.

Eye-defects : This patient's refraction (under homatropine) was corrected by $+0.50^{\circ} \subset +0.50^{c}$ (axis, 90°) in each eye. At the first examination he showed but 1° of manifest esophoria with an abduction of 8°; but esophorial prisms were given and increased, as fast as he showed any surplus, until he disclosed 12° of latent esophoria, and unconquerable homonymous diplopia when prisms were removed.

Treatment and Results : Without any promises of cure, or even expectation of such a result, I advised a *radical elimination* of the *esophoria* by graduated tenotomies upon both interni. To this step the wife and friends assented, with the hope that some possible improvement might follow the removal of so great a leakage of nerve-force as eye-strain was causing in this patient.

After an operation on each internus I advised the family to give this patient absolute quiet and rest in a family sanitarium near Buffalo for several months. He was not a safe member of any private family at this time, and was too hard to control even for any male nurse to feel safe with him while in his exclusive care.

His eye-tests after the operations were as follow: No

hyperphoria ; no esophoria ; adduction, 30° ; abduction, 6° + ; right sursumduction, 3° ; left sursumduction, 3°.

This patient followed my advice. He improved very rapidly, and was discharged as cured from the sanitarium after a few months of simple rest.

A friend of this patient reported, only a few weeks ago, that the recovery of this patient seemed to be complete and permanent. He has again engaged in business and shows no signs of mental disturbance.

As this patient was only under my care for about ten days, I doubt if all latent esophoria has yet been fully developed and corrected. The wonderful change that immediately followed the work done upon the eye-muscles indicates quite clearly (to my mind) that the cerebral disturbance in this case was dependent largely upon an extreme physical depression produced by eye-strain, and that recovery would have been improbable without the operative work that arrested the leakage of nerve-force in this patient.

CASE II. *Melancholia with Suicidal Tendencies.*—Mr. T., aged 37 ; married ; insurance-broker. Referred to me by Dr. Trask, of Astoria, Long Island, N. Y.

Family History : Mother has been an invalid most of her life. One sister has had nervous prostration. An aunt was insane. An uncle committed suicide. Some phthisis in his paternal ancestry.

History of the Case : This patient has suffered for twenty years with sick headaches and for over two years with insomnia.

During the past two years his relatives have noticed that morbid feelings and delusions have been manifested from time to time, and that gradually the *condition of melancholia with suicidal tendencies* had developed to such an extent as to cause alarm. He had been for months unable to attend to business, and most of the time of late it had been deemed imprudent to leave him without a companion. All medical treatment and a trip to Europe had afforded no permanent relief.

Eye-defects : This patient's refraction (under atropine) was corrected by — 3.50ˢ ◯ — 1.00ʳ (axis, 90°) in each eye. A marked hyperphoria and a decided exophoria also existed.

Treatment and Results : Glasses for constant wear (to

fully correct the refraction) were ordered, and two graduated tenotomies were performed (one on the inferior rectus and one on the right externus).

Within six weeks this patient regained his health and all signs of mental depression disappeared.

He has been seen by me from time to time during the past eight years and no relapse has ever occurred, although he has passed through extreme business anxieties and other experiences that would have prostrated him in the past.

CASE III. *Entire Loss of Mental Faculties, with Insomnia and Symptoms of Cerebral Softening.*—Mr. P., aged 41; married; manufacturer.

The entire history of this very remarkable case has already been given in Chapter V. The author would refer the reader to page 153 for full details of symptoms and treatment.

This patient had gradually reached a point where he could not tell how or where his various garments were designed to be worn; and he had been known to chew on a bolus of food for one hour until told to swallow. He had for months been cared for as a child would be; his bills had been paid for him by his attendant; he had been regarded as a case of hopeless softening of the brain by all physicians that had been employed to treat him, and yet he made a complete recovery by the correction of his eye-defects.

Seven years have passed without a relapse. He is to-day actually engaged in his old business pursuits, that are very taxing to his strength and in which large financial interests are involved.

CASE IV. *Melancholia, with Violent Outbreaks and Delusions.*—Mr. C., aged 47; married; treasurer of a corporation.

Family History: No history of mental disease in ancestry as far as patient knows. Most of his immediate relations and his ancestry are somewhat delicate in physique. Some have headaches. No phthisical tendencies.

History of the Case: This patient was from youth of a delicate build, thin, pale, and not overstrong. He had cultivated literary tastes, wrote poetry, was extremely bright in an intellectual way, and was appointed early in charge of very large financial interests. His temperament was a wonderfully cheerful one, both in his family and his outside social relations.

About three months before I was called to see him he had caused great alarm and anxiety among his friends and family by unmistakable signs of mental failure and delusions. Among the first symptoms that he showed were several startling confessions on successive days to his wife of infidelities with nearly all of her most intimate friends (at times and places which were incompatible with the facts). Soon his mental vagaries became apparent in business, so that he was asked to take a leave of absence for rest and recuperation.

Within the next few weeks a marked melancholia developed. He became reticent, taciturn, and would sit motionless for hours, refusing food, disregarding attempts at conversation, and with a fixed gaze upon some object in the room.

He became too weak to go out of the house often for exercise; in fact, a decided disinclination developed to leave his room, and persuasion was of little avail. Extreme irritability developed whenever he was disturbed.

This was the condition of the patient when I first was called to see him at his country home.

On looking at his face I saw at once that *one eye converged extremely, while the other apparently rolled up underneath the eyelid when shielded.*

Binocular vision was evidently difficult for any length of time. He had never worn glasses or had his eyes tested.

His mental condition was too sluggish to get any intelligent answers to questions; so, at my suggestion, he was brought to New York and a male attendant was obtained at once.

When brought to my office, some eye-tests were obtained with the greatest difficulty that pointed to extreme astigmatism in the left eye (+ 4.50°; axis. 115°) and in the right eye a less defect (+ 1.00 ⊃ + 1.00°; axis, 75°). His muscular tests could not be accurately made on account of a practical strabismus (mostly latent) that had existed for a life-time, according to his wife's statement.

Treatment and Results: Glasses were ordered for constant wear to correct the refractive error of each eye; but the patient became quite violent, refused to eat for several days, passed his fæces and urine in bed and in his clothing. and in many other ways gave evidence of such extreme weakness that an approaching dissolution was feared.

Eventually, under careful rest, forced feeding, and stimulants. the patient grew sufficiently strong to be brought again to my office. I then decided to *operate on the left internus* at once, without further attempts at accurate measurement of the esophoria. Accordingly as free a button-hole tenotomy as possible was done on the left internus.

Gradually the patient began to grow stronger and less difficult to control. He wore his glasses constantly and was given static electricity, massage, and daily walks for some months.

Within a year his recovery was complete and he returned to his former position, that he fills to-day with credit to himself and to the satisfaction of all.

Within the past year the opportunity to make careful eye-tests upon this patient gave me some very interesting information. Over five years had elapsed since the operation and the prescription of glasses, upon rather doubtful tests; hence it was of great interest to me to note how closely I had approached to orthophoria and the true refractive condition.

I got the following tests: O. D. V., $\frac{2}{3}\frac{0}{0}$ —; O. S. V., $\frac{2}{3}\frac{0}{0}$ —; no esophoria; left hyperphoria, $\frac{1}{4}°$; adduction, 15°; abduction, 4°; sursumduction, 3° in each eye.

These tests would justify a suspicion of some latent esophoria; but the patient was "too busy to remain for further observation."

The complete and permanent recovery of this patient is (to my mind) one of the most remarkable experiences of my professional life, in many ways. The refraction was perfectly estimated by me under most disadvantageous conditions; the operation was performed as a *dernier ressort* without very accurate previous measurements to guide me; the physical condition of the patient had steadily grown worse until the operation, and steadily improved after it was performed; and the patient recovered his health perfectly within a year and exceeded all previous weight by about twenty pounds.

CASE V. *Melancholia with Morbid Impulses, Associated with Great Mental Confusion and Distress and an Obstinate Neuralgic Affection of the Prostate Gland.*—Mr. S., aged 23; single.

Family History: The mother of the patient suffers from

neuralgia and headache. The paternal grandfather had paralysis. The paternal grandmother was "extremely delicate." One brother suffers from headaches. Another brother is very excitable and of a highly nervous temperament. No case of consumption has ever occurred in any branch of the family.

Eye-defects : Hypermetropia (latent), 2.50 D.; esophoria (manifest), 4°. Subsequently 12° were elicited prior to any operative procedure.

History of the Case: This patient had been under medical care for many months for a prostatic neuralgia, and had derived no benefit from local or general treatment. He developed *melancholia*, and would frequently retrace his steps for several blocks, during a stroll, in order to touch some object which he felt he should have touched when he passed it. The use of his eyes intensified his mental symptoms markedly. He also suffered from morbid fears. He had never had venereal disease, nor was he a victim to self-abuse.

The melancholia of this patient had caused the greatest alarm and anxiety to his parents and friends. It became very difficult to arouse him to any physical exercise, and his previous habits of labor, reading, and conversation were so changed as to make his future mental condition seem most critical.

Treatment and Results : After partial tenotomies were performed upon his interni, and his hypermetropia was corrected by + 1.50 spherical glasses, his recovery was very rapid and complete. He has had no abnormal mental symptoms of neuralgia of his prostate since the first operation (now nearly ten years). His father, one brother, and a sister have since been examined by me, and all had very marked eye-defects.

In some respects this is one of the most remarkable cases I have yet observed. The mental condition of the patient, prior to the relief of eye-tension, was such as to justify the worst forebodings. Neither he nor his family had ever suspected any eye-defect in spite of the fact that his latent hypermetropia was of a very high degree (nearly 3.00 D.), and his latent esophoria was of an equally high degree. His prostatic neuralgia was of a severe and intractable type, and its cause could not be discovered; yet it disappeared at once after a free operation upon each internus.

The change in the mental condition of this patient after the relief of his eye-strain manifested itself at once in his desire to assume active employment. He immediately turned his attention to his profession (that of art), in which he soon gained an enviable fame.

CASE VI. *Melancholia, Extreme Nervous Prostration, and Delusions.*—Mr. S., aged 31 years; single; merchant.

Family History: Both parents living. Father has marked eye-defects and has been a victim to sick headaches. Two brothers and a sister have severe " bilious " attacks. No consumptive tendencies in the ancestry.

History of the Case: The following extract from the father's letter prior to my seeing this patient tells its own story :—

" I have a son, now 31 years old, who graduated from college eleven years ago, ranking second in his class in scholarship. When he was 2 years old I took him to Boston, where, on two different occasions, he was operated upon for strabismus. The eyes have never focused alike, and he always uses one eye when he reads or does fine work.

" He is a natural mechanic and worked upon watches before he entered college. After graduation he devoted himself to fine jewelry. He borrowed capital and started in business for himself.

" Soon he began to write the strangest kind of letters. We tried to combat his ideas, as we thought them to be vagaries that he would soon get over. But the next news was that he was insane and that I must go and take care of him.

" I found his goods exposed, his business abandoned, and himself in the police-station under the charge of keepers. His mind seemed to be wrestling with people, orders, numbers, and colors. For weeks he had been seeing imaginary people marching by his store in a certain regular order, and the colors of the dresses, neckties, etc., had to be in a certain sequence. Everything to his mind was fixed as by fate, and he was held tight in its grasp.

" I tried to argue and to get him off from his strange delusions, but his reply was that he had ' to follow the signs.'

" I brought him home and in three weeks (at the advice of two physicians) I had him taken to the Brattleboro Insane-asylum. He was there four months. The doctors pronounced his case to be one of ulceration of the stomach, which they had treated and essentially cured.

"He became so discontented and unhappy that we removed him (two years ago) to our home. The patient would talk of nothing until away from the institution, when he begged of us not to send him back again. He improved at home and tried to resume business again; and now, poor boy, he has again relapsed into his previous condition. He is again 'seeing signs,' and, in spite of all we can say, they are again influencing his actions."

Eye-defects: This patient showed, under atropine: Irregular refraction; O. D. V. made $\frac{20}{50}$ by $+ 2.00^s \subset + 0.50^c$ (axis, 90°); O. S. V. made $\frac{20}{30}$ by $+ 4.00^s \subset + 0.50^c$ (axis, 90°). A very marked jump (on exclusion of either eye) *indicated some exophoria and extreme left hyperphoria;* corrected by a 6° prism. With the phorometer he showed *no hyperphoria,* and *esophoria of 3°.* He suppressed visual images so markedly that his muscular tests were deemed unreliable.

Treatment and Results: Glasses for constant wear were ordered as follows: O. D., $+ 1.00^s \subset + 0.50^c$ (axis, 90°); O. S., $+ 3.00^s \subset + 0.50^c$ (axis, 90°).

He was instructed to go home for three months and to wear his glasses constantly while awake, and to report at the end of that time for further observations respecting his heterophoria.

This patient, on his return home, began at once to improve. He has resumed business and has carried it on successfully. He has totally recovered his mental faculties and improved wonderfully in health and weight.

His mother writes me, about two months after the prescribing of glasses: "My son is so much improved that he suggests that there is no need of my going to New York with him, and we think the same. I feel quite sure now that his brain-trouble proceeded from his eyes."

The heterophoria that doubtless exists in this patient has not yet been investigated by me. What has been gained thus far has been due to the correction of the refraction alone.

This patient visited my office six months after this report, coming to the city alone. He seemed a totally changed man, was buoyant and cheerful, full of vigor, and in perfect mental balance.

He had gained greatly in weight and said that he could not be without his glasses for a moment without great distress in his head and eyes.

CASE VII.—*Complete Nervous Collapse. Suspected Brain-softening.*—Mr. C., age 41; married; accountant. Referred to me on August 31, 1886.

Family History: Father died of pneumonia; mother died of heart disease. All brothers and sisters have headache, inherited from mother, and are rather delicate. Insanity on paternal side. Four maternal aunts died of phthisis. One daughter died of phthisis. One son had infantile paralysis.

History of the Case: This patient has suffered from sick headache since he was 18 years of age, averaging one headache a week. Since 1880 he has averaged two each week.

In 1885 his headache became constant, associated with insomnia, great nervousness and marked debility. He had to give up business, as he was unable to bear the strain.

In May, 1886, he was examined by a prominent physician, *who diagnosed " incipient cerebral softening."* He ordered two months of travel.

He came back with no marked improvement and was sent to me for examination.

When I first saw him he was extremely weak, very emotional, and extremely nervous. His digestion was very poor. He slept very little. The long continuance of his nervous breakdown and the failure of rest and travel to restore health had given him just cause to fear that his life was in peril, and that the brain was the seat of organic disease.

Eye-defects: This patient's vision was $\frac{20}{15}$ in each eye with the following glass: O. D., — $0.75^s \bigcirc — 0.25^c$ (axis, 180°); O. S., — $0.50^s \bigcirc — 1.00^c$ (axis, 175°).

He also had a *high degree of latent hyperphoria and eso-phoria.* These defects were made manifest by the use of prisms, and were relieved, from time to time, by graduated tenotomies as his tests warranted.

The full details of the ocular treatment in this case would require too much space to allow of publication here.

His muscular tests at the present time are as follow: Adduction, 32°; abduction, 6° +; sursumduction, right, 2°; left, 2°; no hyperphoria; no esophoria.

He now uses $+ 2.00^s$ over the above distance glass for reading.

Treatment and Results: This patient was instructed by me, from the beginning of the eye-treatment, to wear constantly the glasses that fully corrected his refractive errors.

As fast as any marked error of adjustment of the eye-muscles was disclosed either prismatic glasses were also worn in combination with his refractive correction or graduated tenotomies were employed to radically correct the defect.

Within a few weeks this patient improved sufficiently to resume his former occupation. Within one year he was perfectly restored to health. Nearly eight years have now passed without any material change in his ocular tests, and his health remains perfect.

CASE VIII.—*Nervous Prostration of Some Years' Duration, with Mental Disturbance.*—Miss W., aged 24. Referred to me on December 6, 1894.

Family History: The father has had a very weak stomach since he was 21 years old. He has a high degree of esophoria. The mother is very well and strong. One brother has " bilious " headaches. Another brother had a complete nervous collapse when fitting for college. The paternal grandfather died of phthisis.

History of the Case: This patient weighs ninety-seven pounds, and has never weighed over one hundred and two pounds. She has had a twitching of her shoulders since she was 15 years old. About three years ago she became very nervous, and began to be a victim to various fears. Then she lost strength, began to feel generally miserable, had pains in her back and other places, lost control of her emotions, and finally had a nervous fever.

Since recovering from her fever she has had a constant pain in her head, some hallucinations, a marked loss of emotional control, great confusion in her head, and extreme mental depression.

One year ago she went to a sanitarium and was treated by electricity and massage; by this treatment she was helped somewhat, especially in the control of her emotions. She was then sent to me from Canada, to see if she could be cured.

Eye-defects: This *patient was emmetropic,* having, under

homatropine, $\frac{20}{15}$ vision in each eye and tolerating no plus glass.

Her muscular tests at the first visit were: Adduction, 21°; abduction. 6°; sursumduction, right, 2°; left. 2°; no hyperphoria; esophoria. 2°.

By the judicious use of prisms *about 14° of latent esophoria* were made manifest, accompanied by homonymous diplopia when the prisms were decreased and by marked comfort when the prisms were worn.

Treatment and Results: The treatment of this case consisted of one graduated tenotomy on each internal rectus.

The eye-tests. one week after the second tenotomy. were: Adduction. 30°; abduction. 10°; esophoria. 1°; no hyperphoria.

The improvement in her condition is best shown by a letter from her, dated January 13, 1896, in which she says:—

MY DEAR DR. RANNEY:

So many times during these past months I have thought of writing to you, and scarcely know why I have not done so. To-day I feel that I am, apparently, a most ungrateful girl, and when your letter came, this morning, was truly ashamed to think that you had written to ask for the report which a grateful heart should have sent you unasked.

As you know, doctor, I very gradually and very wonderfully improved in health during the months between December, 1894, when you operated on my eyes, and May of last year, when I saw you last. Ever since that time I have been going steadily onward and leaving the terrible past behind me. How terrible it was *no one* can ever know. nor what an agony of suffering I have been released from.

Until about two months ago I could not, with comfort, apply myself to any occupation that required close attention.—as reading. writing, sewing, etc. The old distress is, however, almost a thing of the past. and I hope and trust that by taking care I shall ere long be quite well again.

I do feel that my going to you was most providential, and the result has been miraculous in my case. I only wish all suffering ones could be helped as I have been.   Very sincerely yours.

                                    MISS W.

In September, 1896. this patient called on me upon her return from her wedding-trip to Europe. and reported that she was perfectly well, but had some headache after excessive use of her eyes.

Her eye-tests were: Adduction. 31°; abduction, 9°; sursumduction, right. 2°; left. 2°; esophoria. ¼°; no hyperphoria.

CASE IX. *Nervous Prostration, with Marked Mental Disturbances that were Attributed to Sun-stroke.*—Mr. W., aged 35; married; merchant.

Family History: One brother has severe headaches. The mother is living (62 years old) and is well and strong. The father died years ago, but had sick headaches during his life. No hereditary tendencies to nervous diseases.

History of the Case: Several years before this patient was placed under my care he was prostrated by the heat while out-of-doors. He was not unconscious, but was made temporarily ill.

To this accident most of the subsequent nervous phenomena exhibited by the patient have been attributed by himself and his immediate family. The restoration of a relative to health by eye-treatment after nervous prostration brought him to consult me.

For five or six years prior to my examination this patient had suffered more or less constantly with distressing feelings of congestion in the head; very marked confusion of thought, at times; great mental depression; a marked loss of emotional control; a gradual loss of physical strength; considerable gastric and intestinal disturbance at times; and a decided loss of the fine business ability that he had formerly manifested.

The condition of this patient had caused marked anxiety and alarm in his immediate family and among his intimate friends. The capitalist, who furnished him money for his business, consulted me as to the withdrawal of his capital, because he had cause to be distrustful of the business ability of the patient and he feared the development of some hopeless mental disease.

Although the patient had never himself suffered from any trouble with his eyes, I found on questioning him that he had spent many hours each day upon his ledger, and that any excessive eye-work almost always aggravated his symptoms. He had tried long periods of rest from his business with some benefit, but a return to his labor had always been followed by a relapse of his old symptoms, which were gradually increasing year by year. When he came to me for advice he was extremely depressed and had suggested to his partner that he had better retire from the firm, although this meant his financial ruin.

He *believed himself to be a victim to sun-stroke*, and doubted if restoration to health was possible by any treatment. He had never worn glasses.

Eye-defects: On testing the eyes of this patient the following conditions were found:—

O. D. V., $\frac{20}{20}$ —, made $\frac{20}{15}$ by $+$ 0.75^c (axis, 90°).

O. S. V., $\frac{20}{20}$ —, made $\frac{20}{15}$ by $+$ 0.75^c (axis, 85°).

Under atropine this patient required, in addition to his astigmatism, $+$ 1.50^s to give him normal vision. He was given for constant wear a *full correction of his astigmatism*, combined with $+$ 0.75^s in spectacle-frames.

Treatment and Results: He showed, before and after these glasses were ordered, a high degree of esophoria. To correct this *two graduated tenotomies were performed upon his interni.* He showed, after these operations: No esophoria; no hyperphoria; abduction, 8°.

He immediately became a changed man. His confusion of thought, despondency, and loss of emotional control ceased. He again regained his old business ability and has never lost it during the past ten years.

Within a year this patient was seen by me and reported that he *continued to be absolutely well* and had never had any return of his old symptoms since the eye-treatment.

CASES X AND XI. *Complete Nervous Prostration of Several Years' Duration.*—In connection with the treatment of headache discussed under Chapter III. the history of Case XV is given, and also one other case is reported, in connection with it, as nearly identical. These two histories might be reproduced here as bearing also upon the relief of nervous prostration, as well as headache.

One of these cases had worn a plaster jacket for one entire year while under the care of an eminent orthopædist, while the other had been confined to the house for several years and was carried into my consultation-room when I first saw her.

I regard these two cases as quite remarkable recoveries, after all the hope of cure had been practically abandoned. *Eye-treatment alone yielded results that no other form of treatment had afforded.* The curative effects seem to be permanent.

The first of these two patients visited my office only last week ; the other was employed, when I last heard from her, as a teacher of physical culture in a female college.

CASE XII. *Complete Nervous Prostration. Failure of Rest Cure.*—Mrs. W., aged 40 ; has three living children. All the children have been delicate. [Within the past eight years marked eye-defects have been corrected by me in all of these children.]

Family History : Mother died of consumption. Father died of softening of the brain. One sister died of uræmic convulsions.

History of the Case : This patient was rather delicate during childhood and early womanhood. She had been able to travel extensively, amid luxurious surroundings, and in this way had kept up her physical health. When about 30 years of age, after the birth of her children, she broke down in health and was operated upon for lacerated cervix. No very marked or permanent improvement followed.

She was treated by various methods of treatment during several years, without very marked change in her physical condition. At the solicitation of friends she then spent several months undergoing the "rest cure." The benefits from this treatment were, at first, quite marked, but within a year her old symptoms of nervous prostration returned, and she was advised to have her eyes examined.

Eye-tests : On the first examination patient showed vision of $\frac{20}{20}$ in each eye ; esophoria, 1° ; abduction, 7° ; no hyperphoria ; right and left sursumduction, 3° each.

Under atropine she disclosed latent hypermetropia of 1.50 diopters. Subsequently she disclosed quite a high degree of latent esophoria.

Treatment and Results : This patient was given a full correction for her refraction ($+1.50^\circ$) for constant use. A graduated tenotomy was also performed upon each internus, correcting all latent esophoria and bringing her abduction up to 8°. No more latent esophoria has been disclosed since this operation.

The effects of the eye-treatment were magical. The patient completely regained her health within three months. For over ten years she has remained well without any return of her nervous prostration. During this period she has experienced prolonged and intense anxiety at times. and has also passed through periods of great sorrow. without causing a relapse.

CASE XIII. *Complete Nervous Prostration of Over Five Years' Duration.*—Miss M.. aged 35 years. Referred to me on October 23, 1891, by Prof. J. R. Nilsen, M.D.

Family History: Father living and has a great deal of headache. Mother died from some unknown cause. Severe headache is common on maternal side. One brother, one maternal aunt. and one paternal cousin died of phthisis. Paternal grandfather and paternal aunt died of paralysis.

History of the Case: Patient was not a strong child. She could never use her eyes at night. When a girl at school she used to have very severe headaches, lasting for days and confining her to bed. Ever since then she has had severe headaches. She was obliged to leave college on account of her eyes.

Five years ago she broke down completely with nervous and physical prostration. She had not been able to sit up all day since her breakdown. She has been treated by numerous doctors. She has had her eyes examined by two oculists, who have told her "they would be all right as soon as her general health improved," in spite of her serious errors of refraction that they must have recognized.

She was placed under the care of an eminent gynæcologist in New York City for the investigation of her pelvic condition. He referred her to me later for an opinion relative to her eyes.

Eye-tests: Having been informed of the opinion of other oculists regarding the eye-conditions of this patient prior to my own examination, I was naturally more than surprised when this patient disclosed the following refractive tests: O. D. V., $\frac{20}{\times\times}$—, made $\frac{20}{\times}$ by + 1.50°; O. S. V.. $\frac{20}{200}$, made $\frac{20}{30}$ — by + 3.00° $\bigcirc$ + 0.50° (axis. 120°).

Atropine was at once ordered, as no muscular tests were of any value. until such extreme error of refraction was corrected.

Under atropine her tests were: O. D. V.. $\frac{20}{40}$, made $\frac{20}{30}$ by + 4.50°; O. S. V.. $\frac{20}{40}$. made $\frac{20}{30}$ by + 6.00° $\bigcirc$ + 1.00° (axis, 90°). After the correction of her refraction by glasses (+ 1.00 less than her atropine test) her muscular tests were : Exophoria, 2°; no hyperphoria : adduction, 20°; abduction, 9°; sursumduction, right, 5°; left, 4°. Later the patient disclosed quite a high degree of latent exophoria.

Treatment and Results: The refraction was first corrected, as usual, by the following glasses, given for constant wear: O. D., + 3.50ˢ; O. S., + 5.00ˢ ◯ + 1.00ᶜ (axis, 90°).

Exophorial prisms were then put over the hypermetropic glasses and increased in strength as fast as the tests warranted it. In five days, while wearing 4° prisms, base in, the patient showed: Exophoria, 7°; no hyperphoria; adduction, 20°; abduction, 13°. A graduated tenotomy was then performed upon the left external rectus, and two days later the right externus was also button-holed.

The improvement in this case was very marked from the first. Two weeks after the last operation she reported that she had had no headache for a week; could read with more comfort than ever in her life, and could walk half a mile, whereas two weeks ago she could not walk a block without great fatigue. Her tests at that time were: O. D. V., $\frac{20}{20}$; O. S. V., $\frac{20}{20}$ —; no hyperphoria; exophoria, 1°; adduction, 38°; abduction, 8°; sursumduction, right, 5°; left, 5°.

One month later the patient reported that she could read an hour at night, whereas she had never been able to read at all at night; that her headaches had disappeared; that she slept much better and considered herself about well, except for an attack of the grip. Her eye-tests were the same as in the month previous.

Early in 1896 her headaches returned, and she began to have great difficulty in using her eyes. She accordingly returned to New York in April, 1896, for further treatment. Her distance-glasses were found to be too weak, and they were accordingly increased in strength as follows: O. D., + 4.25ˢ; O. S., + 5.75ˢ ◯ + 1.00ᶜ (axis, 90°). As her accommodation was also found to be very weak, + 1.00ˢ was ordered for use over her distance glasses in reading. Her muscular tests were found to be as follow: No hyperphoria; no exophoria; adduction, 30°; abduction, 7°.

Within the past year she has worn the stronger glasses with great comfort and decided benefit to her general health.

CASE XIV. *Nervous Prostration of Many Years' Standing.* —Miss P., aged 25. Referred to me on September 20, 1892.

Family History: Father has a high degree of hypermetropia, astigmatism, and hyperphoria. Mother has astigmatism, and has suffered from headache and general nervousness. Two brothers had epilepsy; one sister had chorea; two brothers had headache. All brothers and sisters are hypermetropic and astigmatic, and most of them had disclosed heterophoria when I tested their eye-muscles. All the maternal relatives are neurasthenic and suffer from headache. No tendencies to consumption.

History of the Case: From a child this patient had been delicate. She suffered severely from headache, spinal pain, and general nervous and physical prostration. She was physically unable to endure any ordinary exercise, and mentally was so apathetic that she took no interest whatever in the numerous social pleasures and amusements of a luxurious home. For years she had been unable to spend a winter in the North, on account of the weak condition of her lungs. She had consulted numerous physicians both in this country and in Europe.

On her account alone the father had at one time kept his entire family in Europe for two years. During that time she was constantly under the care of distinguished medical men for nervous debility.

The parents of this girl at last determined to bring her back to America and to re-open their home; but she soon became a cause for alarm on account of her physical weakness and general nervous debility. She could endure nothing, could not exercise, was very apathetic and mentally depressed, had imperfect digestion, and looked like a girl going rapidly into a serious physical collapse.

On account of the startling recovery of one of the family after eye-treatment, she was brought to me by her parents to see if anything could be done to check her alarming decline.

Eye-defects: At the first examination this patient's tests were: O. D. V., $\frac{20}{30}$ —, made $\frac{20}{30}$ by — 0.50^c (axis, 140°); O. S. V., $\frac{20}{30}$ —, made $\frac{20}{30}$ by — 0.50^c (axis, 35°); no hyperphoria; esophoria, 9°; adduction, 23°; abduction, 6°; sursumduction, right, 2; left, 3°. She suppressed images, so that a red glass had to be used in some of her tests. Under the full effects of atropine her refraction was: O. D. V., $\frac{20}{70}$ —, made $\frac{20}{30}$ by $+ 1.50^s \bigcirc + 0.50^c$ (axis, 45°); O. S. V., $\frac{20}{100}$, made $\frac{20}{30}$ by $+ 1.50^s \bigcirc + 0.50^c$ (axis, 135°). With this correction she disclosed: Esophoria, 12°.

Treatment and Results: A partial correction for her hypermetropia and astigmatism was at once ordered for constant wear, and, as soon as she would accept it, a full correction for her refraction was given. A tenotomy was performed on each internal rectus.

The improvement began almost immediately after the first tenotomy. She commenced to gain in strength and weight, her apathy disappeared, interest in her home and society returned, and for the first time in years she stayed in New York all winter.

Within a year she was apparently absolutely well, and has continued so ever since. She rides horseback daily; attends to her numerous social duties and amusements; is bright, cheerful, and about twenty pounds heavier than when I first saw her. She has not been obliged to leave New York during a single winter, as her lungs seem strong and all fear of consumption has been removed.

CHAPTER IX.

THE SURGICAL TREATMENT OF ANOMALIES OF THE OCULAR MUSCLES (HETEROPHORIA).

In the preceding chapters I have reported quite a number of cases where, after a correction of the refraction of a patient by glasses had failed to afford relief to the nervous symptoms, the correction of an error of adjustment of the ocular muscles effected a complete cure.

Again, a certain proportion of the cases reported by me in this volume exhibited little or no errors of refraction (even when examined after the instillation of atropine into the eyes).

Some of the cases, moreover, experienced but little benefit from the wearing of prismatic glasses for the heterophoria that existed, although they immediately experienced marked relief after a radical correction of the muscular defects by graduated tenotomies.

I do not think, therefore, that any oculist with a large clinical experience in this special line of work can possibly indorse two deductions that are to-day accepted by some men of note,—viz., that refractive errors always constitute the basis of muscular anomalies in the orbit, or that prismatic glasses accomplish all that operative steps upon the eye-muscles can.

The first of these erroneous conclusions is, to my mind, positively refuted by several emmetropic cases of epilepsy reported by me, as well as by the histories of cases of other types of nervous disturbance given in previous chapters of this volume.

Whenever patients who have no refractive errors make a complete recovery (either by the wearing of prismatic glasses alone or after graduated tenotomies upon some of the ocular muscles) the evidence seems to me conclusive that muscular errors in the orbit are fully as important to recognize and treat as refractive errors. that anomalies of adjustment of the eyes may be independent of any malformations of the eyes themselves, and that severe nervous disturbances may be both induced and aggravated by eye-strain from maladjustment of the muscles of the orbit.

(292)

Again, if a patient, without any error of refraction, experience no benefit from wearing prismatic glasses, properly prescribed, but is relieved at once of nervous symptoms by a radical correction of the existing errors of adjustment of the eyes through a tenotomy, it seems most absurd to me to argue that prismatic glasses are as effective in the treatment of all cases of heterophoria as are tenotomies when properly performed.

I would not be construed as endeavoring to impress the reader with the view that all cases of eye-strain demand operative interference. Such a position is not held by any intelligent and earnest worker in this field, in spite of insinuations to the contrary that are even yet too common in medical discussions and occasionally encountered in print.

Too great stress cannot be laid, in the opinion of the author, upon the importance of an extremely careful correction of all errors of refraction by glasses and also the judicious employment or prescription of prisms (either in combination or alone) before any operative step is even suggested to the patient.

The hints thrown out in Chapter II of this volume should be carefully read and studied in this connection. They will materially aid the reader in determining when the various factors of a complicated eye-problem should be regarded as satisfactorily and positively determined; when the relative power of individual muscles in the orbit justifies and confirms the tests of the phorometer or tropometer; when the type of heterophoria that apparently exists in any case is confirmed by the Maddox rod, the red glass, the jump of either eye on exclusion, and by various other confirmatory tests; and when the amount of heterophoria exhibited by the patient is sufficient to make it imperative or wise to correct it by an operative step rather than by prismatic glasses.

For example, it is not my custom (as a rule) to operate upon either an internal rectus muscle or an external rectus muscle *until the amount of esophoria or exophoria disclosed by the patient exceeds six degrees* (6°). Up to that point I usually employ prismatic glasses to develop latent heterophoria.

Whenever 6° or 7° of esophoria are disclosed by a patient, the externi should not be able to overcome more than 2° of prism, with the base toward the nose, prior to operation. When the same amount of exophoria is disclosed, the externi ought to overcome at least 14° of prism, with base toward the nose, prior to operation.

In either case no hyperphoria should co-exist with the esophoria or exophoria when operative work seems imperative; and the tests shown by the phorometer should be confirmed by the Maddox rod, the red glass, the exclusion-test, etc.

There may be some exceptions to the above statements, but they are sufficiently infrequent to prove the rule. After a very large clinical experience in this field the careful observer may occasionally see some points in a case that might easily escape the observation of a beginner. Under such circumstances operative work on a muscle might, possibly, be deemed wise, in spite of some inconsistencies in the tests disclosed by a patient; but it is not safe for a beginner to rush hastily into operative work upon the ocular muscles.

He should wait until he has studied each case a long time; has tried the effect of prismatic glasses upon the patient; has thoroughly solved and corrected all errors of refraction; and has demonstrated, beyond a doubt, that an operation is clearly indicated and justifies a hope of relief.

Much harm may be done by graduated tenotomies when injudiciously performed upon the recti muscles of the orbit, and it is always wise for the oculist to study each case with the greatest care before deciding as to what muscle is particularly at fault.

In my experience, it is usually wise to correct the hyperphoria first, in cases of hyperesophoria or hyperexophoria,[1] and then to wait for some weeks (if possible) after the operation for the purpose of determining the effect of correcting the hyperphoria alone upon the esophoria or exophoria that was originally disclosed by the patient.

In the light of our present knowledge the *scientific treatment of actual strabismus is becoming one of the most difficult problems of modern surgery;* and the same remark applies with equal force to many cases of complicated heterophoria.

No longer is it considered good surgery to freely divide the conjunctiva, to loosen up the cellular tissue of the orbit, to pick up a muscle on a strabismus-hook, and then to snip it off with the scissors whenever a cross-eyed patient wants the eyes straightened.

<hr>

[1] See page 35.

There is much to be done before the problems of any apparent strabismus are sufficiently solved in each individual case to warrant operative interference. The refraction of each patient should first be carefully determined and corrected for some weeks by glasses properly prescribed; the diplopia of the patient must be frequently and patiently investigated (with a red glass over either eye alternately) at all points and in various positions of the head; the tropometer tests should be carefully noted for a sufficient number of sittings to make them conclusive and reliable; finally, the character of the operation, the muscles which need some modification of attachment, the intervals between the successive operative steps, and various other details must, of necessity, be left to the judgment and experience of the operator.

The greater the clinical experience, the more matured the judgment; and, the better the technic of the operator, the more satisfactory is the result.

Before discussing in detail the technic of performing a graduated tenotomy upon any of the recti muscles of the orbit, it may be beneficial to the reader to become even more forcibly impressed with a few general cautions relating to this work than to allow my remarks to rest at this point. Those which I would suggest here admit of occasional exceptions (like most rules of procedure). They may possibly restrain a beginner, however, from hastily rushing into operations upon the eye-muscles before all the preparatory and precautionary steps have been taken toward a proper solution of each individual eye-problem.

The following points have been of value to myself at many times. and are daily confirmed in my office-work :—

1. *Don't think every case is an operative case,* even if manifest heterophoria is disclosed at the first examination of the patient.

2. *Don't fail to always determine the refractive errors of each patient early and with care.* Employ homatropine or atropine to do so in most cases; and prescribe the correcting glasses for constant wear.[1]

In prescribing glasses I usually *correct all existing astigmatism,* in myopic patients I give the weakest glass that will give fair distant vision. while in hypermetropic patients I try to

---

[1] If astigmatism or hypermetropia is sufficiently important to be corrected, the glasses should be worn at all times. It is not enough to prescribe them for reading, sewing, study, etc., if nervous derangements are to be treated by the methods advocated in this volume.

correct all the hypermetropia in excess of one diopter (even if I have to instill atropine into the eyes at intervals for a few weeks to arrest ciliary spasm).

Be careful that the glasses prescribed are properly centered to the pupils of the patient, and that they are worn without tilting them on the nose and constantly.

3. *Don't fail to watch, during several successive examinations at intervals of a day or two apart, the muscular tests of each patient that is wearing glasses prescribed.* See if the glasses prescribed tend to bring the muscular tests nearer to the standard of orthophoria[1] without any aid from prisms.

I have seen many patients, who at first showed extreme manifest heterophoria, swing into almost perfect ocular adjustment within a few weeks after wearing properly prescribed glasses constantly. Such patients would have been injured by undue haste in operative work. While an operation might have been performed by a beginner on the basis of the first records of the case that seemed to warrant such a step, the results would surely have been regretted later by operator and patient.

Within the past month a patient that showed me unconquerable diplopia (hyperexotropia) at the first examination has been made a case of absolute orthophoria by correcting extreme inequalities of refraction in the two eyes by glasses. She had been regarded by all who had previously seen her as a complicated case of strabismus that operation alone could relieve.

4. *Don't operate until the patient shows consistent tests.* By this I mean that the externi should be markedly below their normal power (8°) if esophoria is to be corrected, and markedly above their normal power if exophoria is to be corrected. When hyperphoria is to be corrected, the difference between the right and left sursumduction ought to be proportionate to the amount of hyperphoria disclosed by the patient.

Genuine esophoria with a high abduction is rare ; and exophoria with a low abduction is always open to suspicion and warrants further investigation.

5. *Don't attempt to guess at refraction by means of the ophthalmoscope, retinoscope, or Javal's ophthalmometer,* to the exclusion of a surer method,—viz., the use of homatropine or atropine and trial-lenses.

---

[1] See page 35.

It is always well to confirm the refractive tests by Javal's instrument, when astigmatism seems to exist, prior to or after the pupillary dilatation.

I once saw three and one-half diopters of latent hypermetropia unrecognized by one of the best ophthalmoscopists of his day. Proper glasses cured the patient, after passing into another's hands.

6. *Don't regard any eye-problem as simple.* The most serious nervous diseases may exist as the result of latent heterophoria, and the solution of the problem may take time and patience.

One of the most remarkable recoveries from epilepsy that I have ever encountered (Case XIII of Chapter VII) occurred in a lady that would have passed the examination of any oculist when I first tested her eyes. She had absolute emmetropia and apparent orthophoria. It took me at least one week to determine what step to take in the treatment of her case.

7. *Don't think that the eye-muscles are properly adjusted because the eyes of a patient can follow any object moved before the face in all possible directions.*

Ridiculous as such tests may seem to some, I have known of positive opinions being given by oculists of note upon information of this character.

8. *Don't fail to employ the phorometer, the tropometer, the perimeter, the Maddox rod, and the red glass* before you come to any final opinion regarding the question of existing heterophoria.

The reader may not have all of these instruments, and in many cases much may be learned without them; but it is not safe to regard any examination of the eyes as complete or final without them, or to be hasty in coming to a decision even after employing them.

9. *Don't fail to study (before operating) the movements of the eyes behind the screen or card placed in front of the eyes alternately,* when the patient is gazing at a candle-flame placed twenty feet from the eyes. Any jump of the eye (when uncovered) is often a valuable hint as to subsequent operative work (see page 36).

10. *Don't fail to study (before operating) the facial expression of each patient, and any peculiar attitudes of the head that may be apparent to a keen observer.*

The attitudes of hyperphoria, anaphoria, and kataphoria are very diagnostic, while the facial expression caused by esophoria and exophoria are easily detected by an expert (see page 65).

11. *Don't fail to study, from the first tests of any patient, the actual power of abduction, adduction, and sursumduction ;* also the arc of rotation (of each eye independently) upward, downward, inward, and outward (as measured by the tropometer).

A *persistently low abduction* always justifies a suspicion of latent esophoria,—one of the most prevalent causes of reflex nervous disturbances.

An *extremely high abduction* points toward a possible exophoria or a latent hyperphoria.

Any marked and persistent defects in rotation (as indicated by the tropometer) are extremely valuable aids in this connection.

12. *Don't be too hasty in diagnosing exophoria and operating upon the externi.* In my experience, uncomplicated exophoria is much less frequently encountered than most oculists seem to think. It is rather an infrequent cause of severe nervous disturbances. outside of headache.

In the twenty-six cases of epilepsy reported by me, only two had exophoria.

13. *Don't make muscular tests without the glasses that have been prescribed to correct errors of refraction.* except for some good reason. such as a high degree of myopia, ill-fitting frames. etc.

It is occasionally wise to remove the glasses of the patient (while trying the muscles) when there is a suspicion that a tilting of the frame is causing a slight amount of hyperphoria, and sometimes for other reasons too numerous to mention here ; but it is not the surest way to make muscular tests upon a patient whose refractive errors are sufficient to warrant the prescription for glasses.

14. *Don't fail to appreciate the difficulties of testing or treating highly-myopic patients for heterophoria.*

A strong biconcave glass. when decentered (as they too commonly are in myopes), acts practically as a strong prism would do if placed before the eye. The removal of the myopic glasses, moreover, makes the patient often too blind to see the

test-flame with distinctness. For these reasons it is often a difficult matter to be certain that the tests of a highly-myopic
patient are as reliable as those made upon others who are
emmetropic, or who wear a lighter glass that is less apt to sag
and become decentered.

15. *Don't expect or promise marked relief of nervous disturbances from prismatic glasses.*

There are many patients who welcome a prismatic glass as
a great boon, in which case the benefits of eye-treatment are
apt to become apparent to them almost immediately. But, on
the other hand, many patients are very seriously inconvenienced
by the wearing of prisms, and fail to perceive any benefit from
them.

It is my custom to explain this fact to patients and to impress upon them the utility of prisms as a means of diagnosis
rather than of expected relief. It is wise to caution them about
uncertainty in walking, etc., when they are first worn. One of
my patients injured himself quite severely, while trying to step
into his carriage, after putting prismatic glasses on, from a
failure on my part to caution him about making missteps, etc.

16. *Don't fail to keep prismatic glasses on while testing the
muscles* (if they are being worn by a patient for the purpose of
detecting latent heterophoria). The amount of prism worn by
the patient should be allowed for (being added or subtracted, as
the case may be) before the strength of any muscle is recorded.

While any patient is learning (by the aid of prismatic
glasses) to abandon former habits of adjustment of the eyes for
heterophoria, the taking off of the prisms (even while testing
the muscles) forces the patient at once to use the old tricks of
adjustment, and thus to simulate a smaller degree of heterophoria
than actually exists.

17. *Don't keep prisms upon any patient unless the tests
show a distinct surplus of the original heterophoria that the
prisms fail to correct.*

By this caution I mean that no one should force any
patient to accept and adjust for a prism. The tests of the
patient should always show that the prism prescribed corrects
only a part of the heterophoria disclosed (see page 55).

18. *Don't put patients to the expense and trouble of having
prisms* (when employed for purely diagnostic purposes) *combined
with their refractive correction.*

Employ the interchangeable prisms in special frames as a part of the office-equipment (see page 55), for these can be worn over the other glasses and changed in strength as the tests indicate, without delay, inconvenience, or expense.

Spheroprisms or cylindroprisms are quite expensive to buy and discard as often as prisms are altered in strength while latent heterophoria is being investigated.

19. *Don't fail to examine and neutralize every glass prescribed before the patient wears it, and to see if the glasses are properly centered to the pupils.*

I have not for years allowed a glass to be worn by any patient of mine that has not been passed upon by myself. Mistakes are too common, with the best of opticians. to make an oversight in this regard a trivial matter.

20. *Don't fail to remember that patients who either have strabismus or an amount of heterophoria that is on the borderline of squint generally suppress visual images* (either as a constant habit or at intervals).

This fact makes it very difficult, at times, to get the muscular tests of such patients. The red glass comes into use as a valuable adjunct in such cases. By coloring one visual image, it is much easier to teach such a patient to sustain binocular vision and realize the diplopia that may actually exist and yet be unsuspected by the patient and too often by the oculist.

It often takes much time. ingenuity, and patience to teach a strabismic patient to look with both eyes simultaneously.

The steps that may be taken to detect suppression of images by a patient and to teach him to overcome it often require persistent effort on the part of the oculist and quite a large experience in handling such cases. It is often necessary to put quite strong prisms before the eyes of the patient, to color the image, to exclude the vision of either eye alternately for an instant, to get tropometer tests as an aid, and to repeat the lesson many times before any positive records of the existing heterophoria can be made.

21. *Don't be disappointed if some cases of heterophoria show tests that are inconsistent and contradictory.*

Whenever you encounter cases that have esophoria, or homonymous diplopia, at one point and exophoria, or crossed diplopia, at another point ; when eyes swing inward to a marked degree sometimes and outward to a marked degree at others; when the phorometer and the Maddox rod give results directly opposed to the jump of the eyes on using the exclusion-tests; when the patient is not benefited by any prism or refractive glass, then go slowly, use the tropometer, and ask counsel of some one who has a larger experience in such cases than yourself.

22. Finally, *don't be prejudiced.* *Be honest, conscientious, painstaking, and persistent after you are sufficiently in earnest to attempt to master this field of scientific investigation.*

Regarding the actual steps of a graduated tenotomy upon any of the recti muscles of the orbit, little can be added by me, I fear, to what has been already published, except my personal indorsement. The operation which I usually employ is that devised by Dr. George T. Stevens. I regard it as a great step in advance of tenotomies performed by the older methods, in all of which the muscles were drawn out of the wound on a strabismus-hook (after making a large slit in the conjunctiva and loosening up the cellular tissue of the orbit) and then dividing the tendon either completely or on its borders.

By the new method the axis of traction of the muscle is not disturbed; the cellular tissue of the orbit remains intact; there is no conjunctival scar or deformity left; the danger of sepsis is practically avoided; no dressings to the eye and no confinement of the patient are necessary; and the patient is able to get relief without pain (through cocaine) and with little, if any, inconvenience.

The technic of this operation seems to be very simple, but the skill of the operator is shown by the results which he obtains in making the button-hole in the tendon correctly and of sufficient size to allow of the proper extent of slipping of the tendon upon the sclera before it makes a new attachment. In other words, to correct a given number of degrees of heterophoria with certainty requires natural aptitude, acquired skill, and a large experience.

In advancement operations surgical skill of the highest order is particularly demanded. In my general surgical experience of many years I recall no operation that is more delicate and precise.

STEPS OF A GRADUATED TENOTOMY.

1. Instill three or four times into the eye 2 or 3 drops of a fresh 6-per-cent. solution of cocaine, at intervals of about three minutes apart, until anæsthesia is produced.

2. Insert the eye-speculum, without bruising the lids or using force, in case the inferior rectus, the internal rectus, or external rectus is to be operated upon. Some prefer a speculum with tortoise-shell lid-support, but I have found an ordinary silver-wire speculum to be satisfactory, painless, and secure, when properly inserted.

If the superior rectus is to be operated upon the lid is elevated by a retractor with a handle attached. This is held by the assistant.

3. The assistant stands near the chair with some absorbent-cotton balls or a few very fine, small sponges. immersed in a

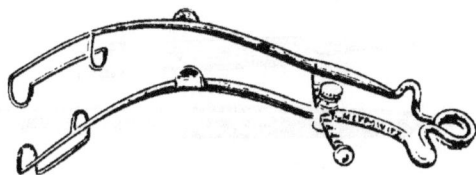

FIG. 36.—EYE-SPECULUM.

weak solution of boric acid or some non-irritating disinfecting solution.

4. The patient is instructed to steadily fix the vision upon some designated point. The point is selected so as to put the muscle to be operated upon on the stretch and to bring it clearly into view. The patient is instructed not to look elsewhere, under any circumstances, during the operation.

5. The conjunctiva is then picked up with a very delicate tooth-forceps and a very small opening is made through it directly over the center of the attachment to the sclera of the muscle that is to be operated upon.

6. This opening is then stretched by inserting the forceps, when closed, into the opening and allowing the blades to spring slightly apart. The elastic conjunctiva contracts later, and leaves only a point for union not much larger than a pin-head.

7. The tendon is then grasped (through the conjunctival opening) slightly back of the exact center of its attachment to

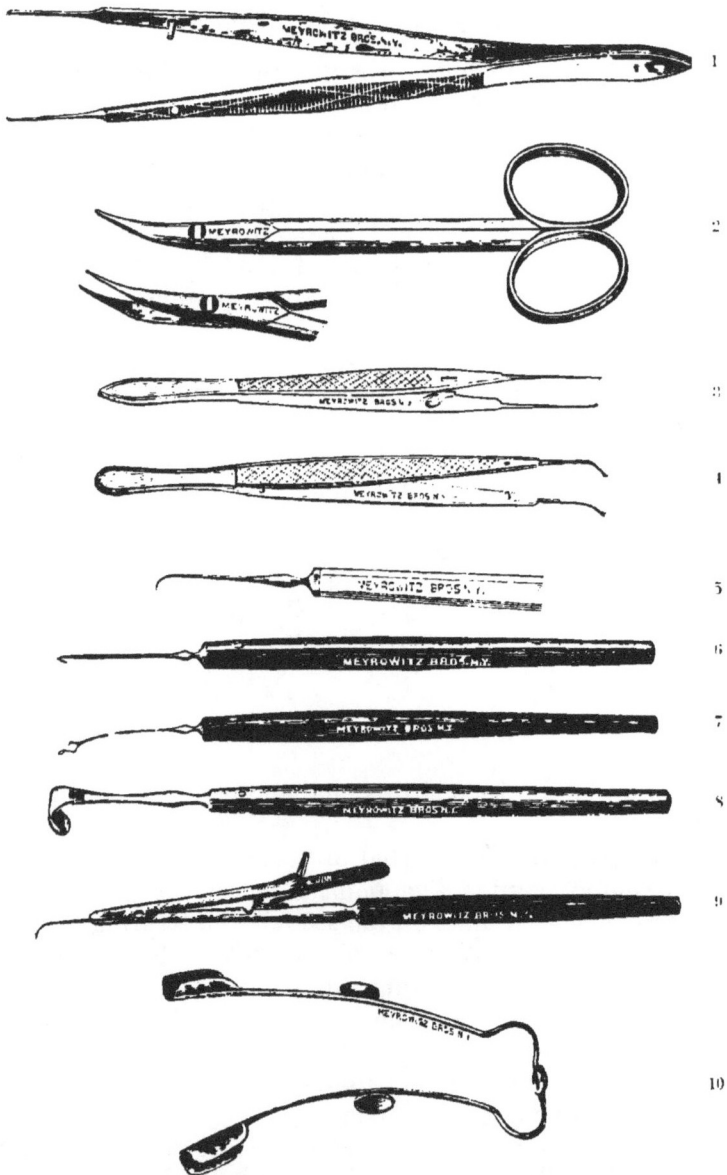

FIG. 37.—EYE-INSTRUMENTS DEVISED FOR WORK UPON THE EYE-MUSCLES BY DR. GEORGE T. STEVENS.

From above downward these instruments are (1) tooth-forceps, (2) Stevens's scissors with very fine points, (3) fine fixation forceps, (4) very fine tooth-forceps, (5) delicate hook, (6) delicate retraction-hook, (7) probe-pointed, double-edged knife, (8) retractor for upper lid, (9) Stevens's needle-holder, and (10) eye speculum with tortoise-shell lid-supports.

the sclera and a snip is made through it, at its central point,
with the tenotomy-scissors.   From this central point the division
of the tendon is then carefully extended, in both directions, as
close to its scleral attachment as possible, until a button-hole of
sufficient size to meet the requirements of each case is made.

The extension of the button-hole is made either with the
Stevens scissors (which have extremely fine tips) or with a very
fine, probe-pointed knife, each of which can be slipped easily
beneath the tendon as far as desired.

Whenever an extreme stretch is desired, the button-hole
may be extended far enough so that the fibers of the capsule of
Tenon alone hold the tendon in its proper relationship to the
eyeball.

In most cases there is practically no hæmorrhage; but in
some the subconjunctival hæmorrhage may be quite severe.   It
is readily absorbed, however, usually within ten days, without
stain or disfigurement.

I employ ice to the wound if the hæmorrhage seems likely
to be excessive.

8. The patient is now allowed to bathe the eye in cold
water for about fifteen minutes; after which the tests are taken
by the phorometer and the manifest results of the operation are
recorded.

9. It is my custom to have my patients bathe the eye
(through an eye-cup) with extract of hamamelis and water in
equal parts at frequent intervals for several hours after an operation; to instruct them not to use the eyes for any close work
for twenty-four hours at least; to keep the wounded eye closed
if out-of-doors; and to have them visit my office daily for one
week in order to test the eye-muscles in their new adjustment.

10. On the third day after the operation it is my custom to
begin daily exercise by prisms of the muscle operated upon;
being careful not to overtax the muscle until the new scleral
attachment is firm and permanent.   Such daily exercise is kept
up until the muscle operated upon shows satisfactory power.

STEPS OF AN ADVANCEMENT OPERATION UPON THE RECTI MUSCLES OF THE ORBIT.

In some cases it is advisable to shorten the tendon of one
of the straight muscles of the orbit.   This may be done (as a

sequel to a button-hole tenotomy) upon the antagonistic muscle when a greater effect is desired, as a primary step in some instances, or simultaneous with a tenotomy of the antagonistic muscle in some strabismic cases. This step is called an "advancement operation."

There are various ways of performing this operation. These are described in modern text-books on ophthalmology, chiefly as a means of rectification of the deformity of strabismus. In most of these the conjunctival incision has to be extremely large, and the entire tendon is exposed prior to the passing of the sutures. For that reason alone, as well as for many others, I prefer in most cases of heterophoria (where a few degrees of effect is wanted) to perform a much simpler operation devised by Dr. George T. Stevens. It seems to accomplish all that can be desired, in most cases of this type, without as large a wound and with greater ease. The only possible drawback that I have found to its employment (when great effects are wished for) is that the single suture is apt to cut through the tendon or sclera from excessive tension; yet, on the other hand, there is no subsequent deformity caused by a wrinkling of the tendon from a puckering effect of multiple sutures or a complicated stitch of one suture.

The steps of the advancement operation devised by Dr. George T. Stevens are as follow:—

1. The eye is cocainized and the conjunctiva opened and stretched in exactly the same manner as described on page 302.

2. The tendon is seized with a fixation forceps at the exact center of its attachments to the sclera (D in Fig. 38) and a free button-hole tenotomy is performed (C to B). The forceps remains attached, and is a guide later to the exact center of the tendon. In some instances it is well to run a silk loop through the tendon to take the place of the forceps.

3. The tendon is now drawn well through the conjunctival opening, the connective tissue between it and the conjunctiva being carefully loosened as the tendon is drawn out. The tendon is controlled by the operator either by means of the fixation forceps in its original position or by the silk loop that has been substituted for it.

4. When the tendon has been drawn forward sufficiently to reach the point (A) desired for a new scleral attachment, it is seized at that point (A in Fig. 38) by a fixation forceps. A

20

V-shaped piece ($A\ B\ C$) is now cut out of the tendon with the apex toward the new attachment of the fixation forceps and its base toward the divided scleral attachment of the tendon.

5. A moderately fine, but strong, suture (either of silk or aseptic catgut) is now passed sufficiently behind the fixation forceps at the point A, to insure a good hold upon the tendon, and the needle is then passed through the conjunctiva and some of the sclera at the point D. The points D and A are then brought closely together by the suture, and the suture fastened securely. The tendon is thus shortened an amount equivalent to the distance between A and B.

6. It is wise to test the patient with the phorometer immediately after the operation; also, to endeavor to obtain a manifest excess of some degrees over the actual amount of advancement required. In my experience, several degrees of effect are frequently lost within forty-eight hours, either by stretching

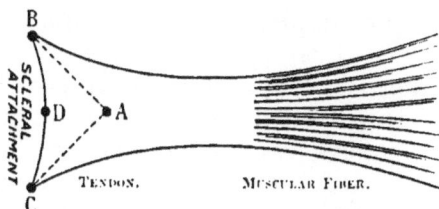

FIG. 38.—A DIAGRAM TO SHOW THE STEPS OF A SINGLE-SUTURE ADVANCEMENT OPERATION.

or partial cutting of the suture, before a firm and permanent union is attained.

7. It is frequently my custom, after an advancement operation, to cover the eye with a shield for at least twenty-four hours, in order to arrest, as far as possible, all efforts of the patient to adjust the eye with its fellow until some union has occurred. Whenever the patient shows much overcorrection, this precaution is not employed, as some stretching is desired.

8. There is, as a rule, very little irritation from the stitch left in the wound, although it is best for the patient to bathe the eye frequently for forty-eight hours with a saturated solution of boric acid in an eye-cup. If the case is progressing favorably, it is my custom not to remove the suture (if of silk) until the third or fourth day. If an aseptic catgut suture be used, the ends may be cut short and the suture will be absorbed.

CHAPTER X.

A FEW PRACTICAL HINTS RELATIVE TO EYE-STRAIN AS A CAUSE OF ABNORMAL EYE-CONDITIONS.

In the preceding chapters I have discussed only from a clinical stand-point the effects of a leakage of nerve-force (caused by eye-strain) upon the nervous system.

It remains for me to say something, in closing, concerning the direct effects of eye-strain upon the organs of vision.

At first thought, this would naturally seem to the general reader to be the most frequent, and perhaps the most important result of imperfect construction or adjustment of the eyes. In answer to this impression, however, it should be stated that, while the frequency of local disturbance may, and possibly does, equal or exceed that of nervous disturbance from eye-strain, the results are far less distressing to the patient, as a rule, and the cause is far more frequently recognized and removed.

A sufferer, for example, from styes, chronic conjunctivitis, chronically inflamed lids, spasm of the eyelids, occasional ulcers of the cornea, flowing of tears over the cheek, punctate hæmorrhages of the eyeball, and periodical attacks of episcleritis, not only suffers less than victims to some intractable and severe nervous derangement, but naturally appreciates the cause of the trouble and seeks relief at the hands of the oculist much earlier than does the victim to reflex conditions.

It often takes many years of suffering and of costly experimentation with drugs, travel, rest, withdrawal from business, etc., to teach those victims (who eventually get eye-treatment for headache, neuralgia, sleeplessness, St. Vitus's dance, epilepsy, nervous prostration, and other reflex nervous conditions) that the cause of their sufferings might have been determined and relieved long before it was.

Again, the sufferer from eye-disease, when he applies to an oculist for aid, always has the support of the family, the family doctor, and all interested friends in persisting in the treatment. He gets advice early, as a rule, from the proper medical source,

(307)

and he is encouraged to keep up his efforts to obtain relief until a cure is effected. Not so, however, with the victim to reflex nervous disturbances! He is too often opposed, when eye-treatment is suggested as a last resort; he is frequently alarmed, without reason, and weakened in his purpose by friends, and even the family doctor, after eye-treatment is begun; and, if he gets relief, he does so under many such conditions of tribulation and mental worry.

The intimate dependence of common forms of eye disease upon both refractive errors and heterophoria is being to-day more generally recognized among oculists of note than in the past. Even the dependence of cataract upon refractive errors (that should have been corrected in youth) is spoken of by many of the modern ophthalmologists.

It is certain that *astigmatism, hypermetropia, and most of the types of heterophoria tend to increase the blood-supply to the orbits.* Such conditions are usually congenital, and, therefore, make those who have either defective refraction or abnormal ocular adjustment peculiarly liable to chronic inflammatory conditions of the lids, the conjunctiva, the episcleral tissue, and the cornea. Later in life, excessive blood-supply to the orbits, as a chronic condition, may help to induce organic changes in the lens (cataract), in the retina, and in the choroid.

It is hoped that the few hints which I shall give in this chapter may aid some of my readers in the treatment and possible prevention of some common forms of eye-disease that are occasionally persistent and very hard to control, unless the cause is eliminated.

STYE AND ENLARGEMENT OF THE MEIBOMIAN GLANDS.—Few diseases of the lids are more common than recurring styes or swellings of the Meibomian glands. They seem to come without cause, and in some cases are recurring and persistent.

Most of such sufferers, when examined under atropine, are hypermetropic and many are astigmatic. If so, it is very important to put glasses on (for constant wear) to correct all refractive errors. Furthermore quite a large proportion will disclose heterophoria. The investigation of some abnormal adjustment of the ocular muscles in this class, by means of prismatic glasses, may accomplish much in preventing subsequent recurrence of the styes or glandular enlargements.

BLEPHARITIS.—Chronically inflamed eyelids are *almost*

always dependent upon latent hypermetropia or hypermetropic astigmatism.

A full correction of the refraction by glasses, if worn constantly, will usually cure these cases within a short space of time. Should heterophoria exist, in addition to refractive errors, prismatic glasses may often be ordered in combination with the spherical or cylindrical glasses, with still greater relief to the patient.

CHRONIC CONJUNCTIVITIS.—The remarks made in previous paragraphs relative to chronic blepharitis apply with almost equal force to chronic inflammations of the conjunctiva. Proper glasses will do more than purely local treatment by astringents, etc.

CORNEAL ULCERS AND EPISCLERITIS.—The former of these conditions is almost universally treated to-day by instillations of atropine at regular intervals,—the latter too often by internal remedies for supposed gouty tendencies (as well as by atropine in the eye).

My convictions regarding these two diseases are very strong and positive.

I have never yet *seen a case of non-specific recurring corneal ulcers that was not due to hypermetropia (usually latent), astigmatism (usually hypermetropic), or heterophoria.*

It is very common for me to encounter cases of this type in my office that have been treated locally for recurring ulcerations by oculists of repute without any apparent regard to the underlying factors of causation.

One of the most intractable cases of recurring corneal ulcers that I have ever seen revealed, when I first tested the patient, a high degree of latent hypermetropia (for which a glass had never been ordered) and an exophoria that bordered closely upon strabismus. Not one of the many oculists whom she had consulted had ever suggested a glass for constant use or had apparently investigated the heterophoria. She has never had a return of the old trouble during the past ten years of relief from heterophoria and the constant use of glasses.

I could report a large number of similar cases of ulceration of the cornea in detail (if space would permit), where no attempt had apparently been made by oculists of repute to prevent the return of this distressing condition by the correction of either the refraction or heterophoria.

The danger of opacities, that often seriously impair vision, ought alone to make every step seem important to the oculist, which tends to prevent a recurrence of corneal ulcers or to hasten their disappearance.

The fact that *episcleritis often seems to exhibit a decided periodicity has caused it to be regarded by many authors as dependent upon a gouty diathesis.*

While I am not prepared to repudiate this deduction entirely as a basis for the treatment of these cases, I am convinced that the detection and relief of heterophoria is far more important than internal medication.

Over ten years ago a gentleman came to my office who had been treated for attacks of episcleritis every spring for more than the ten previous years by oculists of international repute. These attacks had never been known to disappear inside of six weeks, and had often kept him from his business for three months. He had never passed a year without an attack during this interval. He had always been led to believe that they were dependent upon a gouty diathesis, and had taken antigout remedies for many years, both as a preventive and curative step, without any apparent effect upon the attacks.

He figured up, in my office, the total cost of these attacks to him, directly and indirectly, as a moderate fortune for most men, and he feared, not without reason, that future attacks might lead to his business collapse and possible blindness from the corneal opacities that were even then so seriously embarrassing his vision as to cause him great alarm.

With some difficulty fairly good vision was obtained, in spite of the corneal opacities, by means of crossed cylinders. He disclosed a very high degree of latent hyperphoria within the first week of observation. This was corrected by two graduated tenotomics,—one eye being lowered and the other raised. For the past ten years this patient has taken no drugs, and has never had the slightest return of his old attacks of episcleritis. For the larger portion of this period of complete relief he has had to use his eyes almost constantly upon figures for about six hours each day, and he has not lost one day from business on account of his eyes for over ten years. During the ten years prior to the correction of his hyperphoria, he had averaged at least fifty days of entire disability from eye-inflammations during each year.

I have seen several somewhat similar cases that have been equally instructive to me ; but one such extreme example is as apt an illustration as can be brought forward to show the wisdom of seeking for heterophoria early, and correcting it radically.

PTERYGIUM.—The wing-like growths upon the eyeball, that form in consequence of a hypertrophy of the conjunctiva and subconjunctival tissues, and gradually extend toward the cornea, are not due, in my opinion, as much to exposure, climate, etc., as some authors seem to believe.

The remarks made by me in preceding pages relative to chronic orbital congestion as a result of uncorrected errors of refraction or heterophoria apply with equal directness to many cases of pterygium.

It is important to investigate the refraction and eye-muscles of these cases, as soon as any symptoms of a tendency toward pterygium appear. Unfortunately these underlying causes are too commonly neglected, in many cases of this type, until the deformity entails operative procedure for its removal.

EPIPHORA.—A persistent flowing of the tears over the cheek constitutes one of the most distressing of eye-conditions. It is capable of making life almost a burden, in extreme cases, and it is not infrequently encountered in the offices of the oculist and general practitioner.

I would raise my voice here strongly in protest against probing the nasal duct at once in these cases. It is unnecessary in quite a large proportion of such sufferers, as a permanent cure can often be effected without it and seldom by means of it.

I regard *epiphora as a very important clinical evidence of hyperphoria.*

The following cases illustrate this point very clearly :—

Some six years ago a lady came to me from a Western city who had suffered for many years with a weeping of the left eye that nothing seemed to relieve. She had been under the care of many oculists in different cities; had had her canthus slit up; had had her nasal duct probed three times a week for several months at a time; and had experienced little or no benefit.

The weeping eye was seriously defective in refraction. She had worn glasses for many years that had been judiciously prescribed for her refractive errors. She was referred to me by an oculist in Boston, whose son once had been under my care, with a letter requesting that I examine her eye-muscles for latent heterophoria.

This patient disclosed a very high degree of latent hyperphoria, and a graduated tenotomy was at once performed to bring the eyes to the same level. The excessive secretion of tears ceased almost immediately, and for the past five years she has not had to use her handkerchief upon the cheek or to have her nasal duct probed. A very remarkable change has also taken place in her physical and nervous condition.

Another equally striking case first came under my observation several years ago. A young man applied for relief from a persistent flowing of tears over the cheek, that compelled him to hold his handkerchief beneath one eyelid almost constantly, while awake. A 4° hyperphorial prism arrested this distressing condition within an hour. Subsequently a graduated tenotomy effected a complete cure.

If I deemed it necessary I could bring forward a number of similar cases to prove the point that I advocate here,—i.e., that a large proportion of cases suffering from epiphora are victims of hyperphoria. The condition is rather one of excessive secretion of tears than of obstructed outlet, and the investigation of heterophoria ought, in my opinion, to precede any operative procedure on the nasal duct.

PTOSIS.—A tendency of the upper eyelid to drop is quite frequently observed as a direct result of heterophoria. *That it is not always a result of paralysis is quite clearly demonstrated by clinical facts.*

I have seen quite a large number of patients, who had suffered for years from the disfigurement of a drooping lid, make a rapid and complete recovery after a graduated tenotomy for hyperphoria. In a much smaller proportion of cases, where extreme esophoria or exophoria interferes seriously with binocular vision, a condition that closely resembles ptosis may exist. In these cases the deformity is probably produced by an unconscious habit, on the part of the patient, of endeavoring either to exclude the image of one eye or to avoid the perplexity of persistently maintaining binocular vision.

I recall one illustrative case that made a more forcible impression upon me years ago than it would to-day.

A lady once consulted me for headaches that had been

almost constant for some years. They had been associated with
some marked symptoms of nervous prostration.

I noticed, at the first interview, that the lid of one eye
drooped so as to almost shut the pupil from view, and that, when
it was raised by an extreme effort on the part of the patient, the
eyebrow was markedly distorted. Such an extreme condition
made me fear that a complete paralysis of the lid had sometime
existed, but the patient assured me that she had inherited that
defect from her mother.

An examination of the ocular muscles revealed a high
degree of hyperphoria with very slight defects of refraction.

A graduated tenotomy was performed to correct the hyperphoria. To my astonishment, within a half-hour after the operation a very marked change took place in the position of the
drooping lid. Inside of two weeks both the ptosis and headache had entirely disappeared. They have never returned.

SPASMODIC TWITCHING OF THE LIDS.—This common and
annoying condition of the eyes is *almost invariably due either to
latent hypermetropia or heterophoria.*

It belongs to the choreic type of nervous derangements,
and occasionally it may be a precursor of St. Vitus's dance.

The remarks that I have made in Chapter IV apply with
equal force to the investigations and treatment of blepharo-
spasm.

GLAUCOMA AND CATARACT.—We know, as yet, but little
about the causes of these common eye-diseases. They come on
insidiously and without apparent exciting factors, as a rule.

De Schweinitz, in his late work, says in regard to cataract:
" Investigations show that a large majority of cataractous eyes
are hypermetropic and astigmatic, and that the danger of cataract is increased when the astigmatism is against the rule and
remains uncorrected. The *evident prophylactic measure is the
use of proper glasses.*" [1]

This author also indorses the view that similar remarks
apply to the prophylaxis of glaucoma. He says: "Overuse of
the eyes, especially with improperly corrected refractive error,
has a distinct tendency, by causing ocular congestion, to bring
on glaucoma in an eye predisposed to the disorders by changes
in the ciliary region."

[1] Italics my own.

One clinical fact relating to these cases seems to be over-looked by most of the authors on ophthalmology,—viz., that quite *a large proportion of eyes which are rendered blind by disease show a decided tendency toward strabismic conditions after the sight becomes impaired.*

Had the muscular adjustment of these eyes been normal, the simple loss of sight could not have been followed by an altered adjustment of so extreme a type as to entail a stra-bismic deformity.

Sooner or later, in my opinion, a *causal relationship between heterophoria and the development of cataract or glaucoma must be recognized.* Refractive errors may unquestionably predispose to structural derangements of the various component parts of the organ of vision, but heterophoria can entail a strain upon the individual that may also aid materially in causing a chronic congestion of the vessels of the orbit.

DETACHMENT OF THE RETINA.—This sad misfortune is almost invariably the result either of extreme myopia or a direct injury to the eyeball.

This leads us to the consideration of what obscure causes may tend to aid in the development of " progressive near-sight-edness."

When an elongation of the eyeball in its antero-posterior axis (page 9) is encountered it does not necessarily mean a congenital defect. Many people are born myopic, but in quite a large proportion of near-sighted patients the myopia is devel-oped after birth.

Most oculists have cases on their record-books where a moderate degree of hypermetropia has become quite a high degree of myopia in variable spaces of time. For reasons (often very obscure) the eyeballs of some patients seem to undergo a progressive change in shape ; and. when the myopia becomes very extreme, the retina is particularly liable to become detached and to cause sudden blindness in the eye so affected.

It is needless to discuss here the various theories that are held to explain the development of myopia. It may be well, however, to mention (1) the *inflammatory theory* (Graefe), that attributes it to posterior choroiditis ; (2) the *mechanical theories*, that explain it as a result of compression of the eyeball by the external rectus, superior oblique, etc. ; and (3) the *anatom-ical theory*, that attributes it to the size or shape of the orbit, wide pupillary distance, special race peculiarities, the conformation of the face, etc.

A small proportion of myopes owe their defective vision to changes of curvature in the cornea or the lens, but in these cases detachment of the retina is very seldom encountered, except as a result of traumatism.

I am convinced that heterophoria is a very frequent and commonly neglected factor in causing progressive near-sightedness. The difficulties that are encountered in testing the eye-muscles in myopic subjects have been referred to in the preceding section. The greatest care should be exercised in eliminating from the records of a myope the many errors that are liable to result from the decentering of strong myopic glasses.

RETINITIS AND OPTIC NEURITIS.—Although both of these inflammatory states are commonly associated with organic diseases,—such as syphilis, Bright's disease, embolism, toxic conditions, spinal or cerebral lesions, etc..—still, the underlying factors that predispose toward their development are to-day somewhat obscure.

Sometimes both retinitis and optic neuritis develop in subjects that are free from any apparent cause, being induced by simple overexertion of the eyes, intense light, exposure, and other causes of similar character. There certainly is reason, therefore, to suspect, in such subjects, that some underlying cause existed, prior to the development of the inflammatory conditions, which rendered the eyes peculiarly susceptible to strain or irritation from outside influences.

Within the past fifteen years I have tested many subjects in whom retinitis has developed from overwork, excessive light, and exposure, without any clinical evidence of organic disease. In almost every instance has some marked and unrecognized error of refraction been detected, usually combined with a complicating heterophoria of quite a high degree. So enormous has been the percentage of pre-existing eye-strain in those of my patients who have been afflicted with retinitis and optic neuritis that I have begun to doubt if a careful eradication of heterophoria and latent errors of refraction, before these diseased conditions started, might not have saved the sufferer from the sad misfortune that developed.

Unfortunately, it is impossible to prove that the eradication

of the predisposing causes of diseases has positively prevented
their development; but there is sufficient basis, I think, in the
pages of preceding chapters to warrant a search for errors of
refraction and heterophoria in the young, before the nervous
tone of the child begins to be sapped and the dangers of eye
diseases, as well as nervous affections, become intensified by a
neglect of such precautionary steps.

INDEX.

(317)